Li-ion Batteries

RSC Foundations

How to obtain future titles on publication:
A standing order plan is available for this series. A standing order will
bring delivery of each new volume immediately on publication.

For further information please contact:
Book Sales Department, Royal Society of Chemistry, Thomas Graham
House, Science Park, Milton Road, Cambridge, CB4 0WF, UK
Telephone: +44 (0)1223 420066, Fax: +44 (0)1223 420247
Email: booksales@rsc.org
Visit our website at books.rsc.org

Li-ion Batteries

By

Hitoshi Nakamura

Jinnai Industries Inc., Japan
Email: nakamura@jinnai-inc.co.jp

ROYAL SOCIETY
OF **CHEMISTRY**

RSC Foundations No. 4

Paperback ISBN: 978-1-83707-208-8
PDF ISBN: 978-1-83707-207-1
EPUB ISBN: 978-1-83767-417-6
Print ISSN: 2978-1477
Electronic ISSN: 2977-0084

A catalogue record for this book is available from the British Library

The Royal Society of Chemistry is a charity, registered in England and Wales, Number 207890, and a company incorporated in England by Royal Charter (Registered No. RC000524), registered office: Burlington House, Piccadilly, London W1J 0BA, UK, Telephone: +44 (0) 20 7437 8656.

For further information see our website at www.rsc.org

For general enquiries, please contact books@rsc.org

For EU product safety enquiries, please email books@rsc.org or contact Royal Society of Chemistry Worldwide (Germany) GmbH, Römischer Hof, Unter den Linden 10, 10117 Berlin.

Preface

First things first. I regularly post information on a social networking site called LinkedIn and I am often asked if there is an introductory book on lithium-ion batteries that can help people understand the entire process. I have always commented that, although there are some specialized books in the world, there are no books that explain lithium-ion batteries in a broad and simple manner, so the only way to understand lithium-ion batteries is to read and study related papers, even if it is tedious. So, I thought that some people would be pleased if there was a book that provided a bird's-eye view of the technological features of lithium-ion batteries. Over the past few years or so, I have posted a few articles for LinkedIn members and I have received, surprisingly, many comments from people who found it easier to understand than they had expected. So, I have compiled the contents of 16 posts into this book. The title of this book is *Li-ion Batteries,* whereas the title of the draft on LinkedIn is *The Lithium Ion Battery Readers.* This book is designed to provide a broad and simple overview of lithium-ion batteries for those with a university degree or equivalent. The main purpose is to provide those who aspire to learn about lithium-ion batteries with an intuitive understanding of the technical elements of the configuration of lithium-ion batteries. For this reason, it may use expressions that are not always academically accurate. However, the conclusions reached are intended to be correct. For example, where ions are described, it is stated that "atoms are made up of electrons and protons". To be precise, neutrons are also a component of atoms.

RSC Foundations No. 4
Li-ion Batteries
By Hitoshi Nakamura
© Hitoshi Nakamura 2025
Published by the Royal Society of Chemistry, www.rsc.org

However, neutrons are not necessary to understand lithium-ion batteries, so they are omitted for the purposes of this intuitive understanding. However, there should be no difficulty at all in understanding the operation of a lithium-ion battery. Experts may be bothered by some of the finer points, but I hope you will understand the purpose of this book, which is to give beginners an intuitive overview.

The content of this book is based on my own experiences in development. Of course, I have also referred to external literature, but I have chosen the references with the aim of helping readers move on to more in-depth content after reading this book, rather than as evidence for my arguments. Therefore, there are many relatively old documents. These documents are likely to be the ones that those of you currently working in this field have referred to in the past. As for the latest papers, please search for them and read them yourself. This book and the included references should provide you with the basic knowledge you need to read these newest papers.

Hitoshi Nakamura

Acknowledgements

This book is a revised version of a draft distributed on LinkedIn, with more illustrations and diagrams to make it more accurate. The text has also been made easier to understand. The draft was distributed on the condition that it be peer reviewed and, as a result, a number of useful comments were received. These have been incorporated into the book. The names of those whose comments were actually adopted are given here. I also considered the silence of those from whom I did not receive a review as a sign of approval. This book was completed with the help of everyone who is a follower on LinkedIn.

Honorifics have been omitted for simplicity and I am grateful to all those mentioned for their help.

Jyotishman Pathak provided some valuable photographs for Chapter 4.

Katsumi Tanida asked the manufacturers to provide the photographs for this book.

Hideya Yoshitake provided a valuable photograph of the pack in Chapter 5.

Primix Corporation provided photographs of their valuable equipment in Chapter 5.

Kaido Manufactuaring provided photographs of their valuable equipment in Chapter 6.

Nippon Spindole provided photographs of their valuable equipment in Chapter 5.

Aimex Co. Ltd provided photographs of their valuable equipment in Chapter 5.

RSC Foundations No. 4
Li-ion Batteries
By Hitoshi Nakamura
© Hitoshi Nakamura 2025
Published by the Royal Society of Chemistry, www.rsc.org

Hosen Corporation provided materials on the cell assembly process for Chapter 6.

Sylvester Gomes provided valuable illustrations for Chapter 6.

Volker Presser gave me the opportunity to write this book for the Royal Society of Chemistry.

C. K. Park reviewed an early draft of this book and provided detailed feedback.

Jim McDowall reviewed an early draft of this book and provided detailed feedback.

Mubarak Kazmin reviewed an early draft of this book and provided detailed feedback.

Sylvia Lee, to whom I am indebted, when I was thinking about a Chinese version that never came to be.

Diwaker Bandanwal reviewed an early draft of this book and provided detailed feedback.

Finally, I also wish to thank the following people.

Akira Yoshino. I have never worked directly with him, but I have known him personally for over 25 years. We also used to hold small study sessions together on capacitors. After that, he helped me at various times, including giving me appropriate advice at a crossroads in my life. After he won the Nobel Prize in 2019, it became difficult to speak with him casually, but I have made extensive use of the story he wrote about the development of lithium-ion batteries (Japanese version) when writing this book. I would like to express my gratitude for his friendship over the years.

Masaki Yoshio. I learned a lot about lithium-ion batteries from him. I learned many of the fundamentals described in this book from him. The days I spent developing batteries with him were a great source of nourishment for my research. I received my doctorate under his guidance and I am grateful to him for helping me write this book.

Michio Okamura. I learned a lot about the integrated circuits and electrical circuits that make up battery management system AC/DC converters from him. I am grateful to my former colleagues with whom I shared the tough startup phase.

Hitoshi Fujimatsu. I learned many research methods from him. Thank you.

I would like to thank my colleagues at Yamaha Motor, with whom I worked on lithium-ion battery research.

I cannot put their exact names here for reasons of confidentiality, but I would like to thank my friends H. J. and the people with whom we started a new business venture, who have supported my recent research.

I would like to thank my friends CAO, Wang, Hang, and Zhang, who support my activities.

I would like to thank my friends at RISE.

Thank you to the Royal Society of Chemistry editorial team. They patiently taught me how to write a book as a beginner. I am grateful.

I would like to thank all my LinkedIn followers for always supporting my motivation.

Finally, I have a daughter who is bedridden. I created a small laboratory at home so that I could spend as much time as possible with her and my family to continue my research. I was able to write this book with the support of my daughter and my family, for which I am grateful.

Note on Use of the Term "Absorb"

This book describes the phenomenon of lithium ions being incorporated into the material bulk using the somewhat ambiguous expression that lithium ions are "absorbed". Multiple interpretations still exist regarding the forces by which lithium ions are incorporated into the material bulk and interact with the material's internal "surface". While the term "intercalation" is sometimes used for graphite, the standard material in lithium-ion batteries, rather than "adsorption" or "absorption", this book deliberately employs the term "absorb".

RSC Foundations No. 4
Li-ion Batteries
By Hitoshi Nakamura
© Hitoshi Nakamura 2025
Published by the Royal Society of Chemistry, www.rsc.org

Contents

RSC Foundations No. 4
Li-ion Batteries
By Hitoshi Nakamura
© Hitoshi Nakamura 2025
Published by the Royal Society of Chemistry, www.rsc.org

3 Components and Materials That Make up a Lithium-ion Battery   17

4 Li-ion Battery Manufacture: Preparation of Electrodes   43

5 Li-ion Battery Manufacture: Cell Shape and Assembly Method 60

6 Safety and Degradation Mechanisms of Lithium-ion Batteries: How the Battery Management System Works 74

7 State of Health of Lithium-ion Batteries 85

8 Future Prospects for Lithium-ion Batteries   **107**

1 Lithium-ion Battery Principles – Operating Principles of Lithium-ion Batteries

1.1 In the Beginning is the Electron

All electronic devices are powered by electricity. Electricity is the flow of electrons. When electrons move, they do work, for example, by running a motor. When electrons do work, they lose energy. Once the motor is running, the electrons have less energy than they did before. Charging the battery brings the electrons back to their original high-energy state and makes them usable again.

The change in energy of electrons is similar to the idea of a ball moving (falling) from a high place to a low place. In the reverse of the notion of Newton's apple, the energy (learned as "potential energy" in junior high school) becomes high when a ball in a low place is raised to a higher place. When the ball falls, the potential energy is lowered. If there are obstacles in the ball's path, it may be able to move them. There is no change in the appearance of the ball, but the energy is stored in the form of height. In the same way, electrons store energy by moving to a higher energy location.

In other words, the energy changes depending on the state in which an object is placed. The potential energy stored by a ball is under gravity, whereas the energy of an electron is energy that is stored in the electromagnetic world. Coulomb, a French scientist, defined this

RSC Foundations No. 4
Li-ion Batteries
By Hitoshi Nakamura
© Hitoshi Nakamura 2025
Published by the Royal Society of Chemistry, www.rsc.org

force in an experiment using a precise balance that he invented. This is the famous Coulomb's law (eqn (1.1)).

$$F = kQ_1Q_2/r^2 \tag{1.1}$$

Here, Q_1 and Q_2 are the electromagnetic forces of attraction between substances 1 and 2, and r is the distance between the particles. (Since the particles are spherical and the force exerted by the particles is thought to spread spherically, a spherical coordinate system is used.) In other words, the electromagnetic force of attraction between particles is proportional to the force of attraction between them and inversely proportional to the square of the distance. Newton's law of universal gravitation is also known to be proportional to the force of attraction between each object and inversely proportional to the square of the distance, and so the concepts of the two laws are similar (Figure 1.1).

1.2 Voltage and Energy

While potential energy is expressed in terms of height from the ground (in meters), the energy state of an electron is observed as voltage (measured in volts). When electrons are at high energy, the voltage of the battery rises, and when electrons are at low energy, the voltage

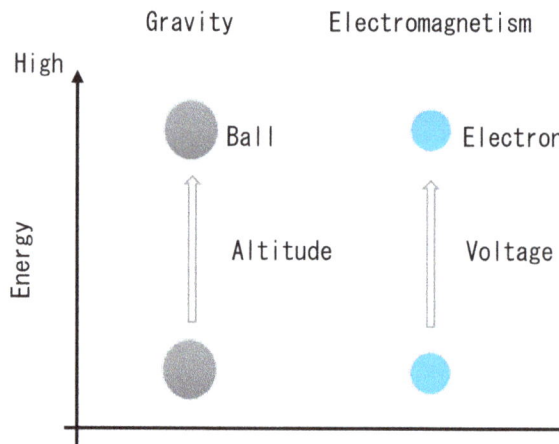

Figure 1.1 Representation of increasing energy under gravity and electromagnetic energy. Note that the image for electricity being stored is incorrect; the correct view is that of the "electron energy" being lifted to a higher position.

of the battery drops. Conversely, if the voltage of the battery is forced upwards by an outside influence, the energy of the electrons moves to a higher state, and the battery can store energy.

1.3 The Relationship Between the Energy Possessed by a Substance and Voltage

In the 19th century, Helmholtz and Gibbs discovered that thermal energy, kinetic energy, the energy of chemical reactions, and electrical energy are all forms of the same energy, and that the law of conservation of energy holds true. They also realized that it is impossible to extract all the energy from a system. They named the maximum amount of energy that can be extracted and utilized from reaction systems, such as steam engines, "free energy". The definition of free energy is given by eqn (1.2).

Free energy = Total energy − Energy that cannot be extracted   (1.2)

Each material constituting a battery possesses its own specific free energy. This energy can be converted into voltage, referred to as potential. The relationship between free energy and potential is shown in eqn (1.3).

$$\Delta G = -nFE_0 \tag{1.3}$$

Here, n is the number of electrons, E_0 is the material's potential (relative to hydrogen's potential), F is Faraday's constant, and ΔG is Gibbs free energy. Gibbs free energy is well-suited for describing changes in free energy at constant volume and is applied to chemically changing systems like batteries. Here, Gibbs free energy can simply be thought of as the maximum energy extractable from the chemical reaction, including the battery. For those wishing to learn more about the origin and concept of free energy, please refer to ref. 1–4.

1.4 Gibbs Free Energy, from the Volta Battery to Lithium-ion Battery

I have explained that the electrical energy possessed by a substance is essentially equivalent to the Gibbs free energy. Everything needs a standard, the difference between the voltage of hydrogen and the voltage specific to the substance concerned is defined as the potential of that substance. The potential of copper is slightly higher than that of hydrogen, at 0.34 V. Conversely, the potential of zinc is slightly lower than that of hydrogen, at −0.76 V. If you immerse these (copper and zinc) in a salt water solution or other electrically conductive liquid and measure the voltage difference, you will observe a voltage of 0.34 V − (−0.76 V) = 1.1 V. This is the famous voltaic cell. If you use gold (+1.5 V) or silver (+0.8 V) instead of copper, you can theoretically make a battery with a higher voltage, but it would be a bit expensive.

The following is based on the literature of Dr Akira Yoshino, who received the Nobel Prize in 2019 for his work on the invention of the lithium-ion battery.[6] Dr J. B. Goodenough, a professor at the University of Oxford at the time, predicted that cobalt oxide containing lithium would have a theoretical high voltage of around +1.5 V with respect to hydrogen, and a Japanese engineer, Dr Mizushima, who was on his team, actually synthesized the compound in 1980.[5] Dr Goodenough published a paper on this invention. It could be predicted that a 4 V-class battery would be created by combining the positive electrode that he developed with lithium, which has the lowest potential of any substance. At the time, disposable batteries using lithium metal were already on the market and organic solvents were used instead of water, as we will see later. Since these were disposable batteries, there were also attempts to make it possible to recharge them. However, rechargeable batteries using metallic lithium were unstable, and there were many cases of fires and other accidents. Dr Akira Yoshino of Japan came up with the idea of combining the materials developed by Dr Goodenough with the carbon-based materials he was researching, and when he tried out the experiment, he was able to charge and discharge the battery very smoothly. Furthermore, the voltage was in the 4 V range, as predicted by Dr Goodenough. This was in 1985. This led to the invention of the lithium-ion battery,[7] and, together with Dr Goodenough and Dr Whittingham, who were the first to propose the operating theory by using TiS_2, this research led to all three receiving the Nobel Prize for Chemistry in 2019.

1.5   Electrons and Ions

The substances around us are made up of atoms. As we learned in junior high school, atoms are made up of a pair of protons (and neutrons) and electrons. Atoms can also be subdivided into electrons and cations (Figure 1.2). In order to use a battery, it needs to be charged. In order to charge a battery, it is first necessary to apply voltage to the

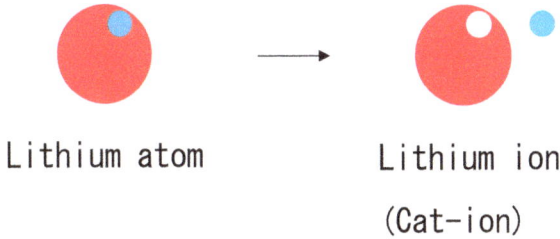

Lithium atom   Lithium ion

(Cat-ion)

Figure 1.2 When an electron is removed from a lithium atom (= metal), it becomes a lithium ion. When an electron is removed, the remaining part becomes positively charged. The positively charged ion is called a cation (cat-ion).

battery and move the electrons around. When the electrons move, the atoms lose electrons and become ions. (Ions that have lost electrons and have a positive charge are called "cations".) By the way, electrons are very small compared to the atomic nucleus at the center of the atom. Of course, the diameter of the atomic nucleus is also very small, but this size is just large enough to be observed using modern technology. However, electrons are smaller than the limit of observation; in other words, they are so small that they are invisible. Interestingly, these tiny electrons have enough power (called electric charge) to balance the much larger atomic nucleus. Remember that electrons are very small compared to ions. This will be an important point from now on.

1.6   The Paths of Electrons and Ions

Because electrons are very small, they can pass through solids, such as metals. On the other hand, ions are much larger than electrons and can move through soft materials, such as liquids, or along fixed paths in solids. When a lithium-ion battery stores or discharges electricity, both electrons and ions must move. Electrons primarily travel through the metal and active material, while ions can move through the gaps created in the active material and through the electrolyte. By extracting the path taken by the electrons, we can use the electricity.

1.7   A Place for Electrons and Ions

You probably know that batteries have positive and negative poles. When a battery is charged, the electrons that were in the positive pole move to the negative pole. When the electrons move, the ions that are

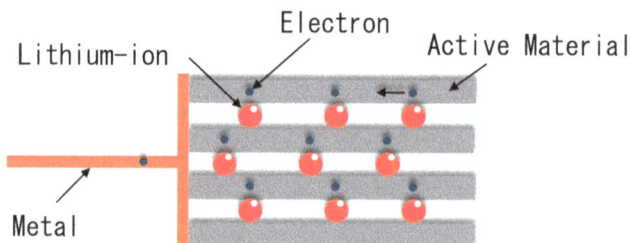

Figure 1.3 Pairs of lithium ions and electrons are contained in a material with a book-like structure.

paired with them also move after them. When a battery is activated, the electrons and ions move around. So, first of all, places and paths for the electrons must be created in the positive and negative poles. Because electrons are very small, they can be stored in spaces that are so small that they can be ignored compared to the ions. On the other hand, ions are much larger than electrons, and so they need a dedicated storage space. It is no exaggeration to say that the main material of a battery is the storage space for ions. I have explained that ions need a solid material with a predetermined path for their movement. In fact, the electrodes that make up a battery are made from materials with such gaps (or paths). If you look at the inside of an electrode, you will see that it has a structure similar to that of a thick book made up of many thin sheets of paper stacked on top of each other.[5] However, the pages of this book are made of a material called a ceramic semiconductor that can conduct electricity. Ions are stored in the gaps between the "sheets of semiconductor" and can move horizontally as needed (Figure 1.3). There are also materials that store ions in spaces other than the gaps between the pages of a book,[5] and I will discuss these in Chapter 3.

1.8 Definitions

1.8.1 Complicated Conventions Related to Cathode, Anode and Positive, Negative Electrodes

When dealing with batteries, you will come across the terms "cathode" and "anode". "Cathode" and "anode" refer to the reduction electrode and oxidation electrode, respectively. These terms were originally defined when electrolysis was carried out. In the world of batteries, the positive pole is called the "cathode" and the negative pole is called the "anode". Those with a good intuition may be wondering why they are called that. This is because the electrodes of a battery switch between the oxidizing electrode and the reducing electrode during charging and discharging. As I will explain later, the definition of oxidation and reduction of a substance is that a substance is oxidized when it loses electrons, and it is reduced when it receives electrons. When a battery is charged, electrons are first lost from the cathode. Therefore, the definition of the positive electrode is the opposite of the definition of the negative electrode when charging. Conversely, when a battery is discharged, electrons return to the positive electrode, so the definition is correct. Before the invention of the secondary battery by Planté, the world only had primary batteries. Initially, batteries only had the function of discharging. In other words, the definition of a primary battery was extended to include secondary batteries.

In lithium-ion batteries, the positive electrode is often referred to as the "cathode" and the negative electrode as the "anode" (this is the convention). This is because the terms were defined based on the state of the battery during discharge. To be honest, when I first started working in the battery industry, I felt a little strange about calling the positive electrode a cathode, but I got used to it. Again, this definition is strictly correct when the battery is discharging, but it is completely the opposite when it is charging. In any case, the battery industry has the following conventions (Table 1.1). I hope this is helpful.

1.8.2 Primary Batteries and Secondary Batteries

A primary battery is a battery that can be used only once, while a secondary battery is a battery that can be recharged and used many times. A primary battery has a high free energy of electrons from the beginning, and when it is inserted into a device, it starts working immediately without the need for recharging. While primary batteries do not have a mechanism to raise the free energy of electrons again after use, secondary batteries can raise the free energy of electrons again by applying voltage from outside. By this definition, a lithium-ion

Table 1.1 A compilation of customary designations. It is complicated, but, for historical reasons, the designations may differ between those who are educated in electrolysis systems and those who are educated in batteries. In this book, these names are used throughout.

	Plus pole		Minus pole	
Battery	Positive	Cathode	Negative	Anode
Electrolysis	Positive	Anode	Negative	Cathode

battery is a secondary battery. In the next chapter I will explain why the free energy of electrons can be raised again in secondary batteries, focusing on lithium-ion "secondary" batteries.

References

1. E. J. Mills, *Proc. R. Soc. London*, 1877, **26**, 504; H. Jahn, *Z. Phys. Chem.*, 1895, **18**, 399.
2. E. Lange and T. Hesse, *Ztschr. Electrochem.*, 1933, **39**, 374.
3. J. M. Sherfey and A. Brenner, *J. Electrochem. Soc.*, 1958, **105**, 665.
4. J. M. Sherfey, *J. Electrochem. Soc.*, 1963, **110**, 213.
5. K. Mizushima, P. C. Jones, P. J. Wiseman and J. B. Goodenough, Li_xCoO_2 (0<x<-1): A new cathode material for batteries of high energy density, *Mater. Res. Bull.*, 1980, **15**, 783.
6. A. Yoshino, *Lithium-ion battery story - in the world of Japanese technology break*, CMC publishing, Japanese edn, 2004.
7. A. Yoshino, K. Sanechika and T. Nakajima, *Jp. Pat.*, 1989293, 1985.

2 Lithium-ion Battery Principles – Basic Operation of Lithium-ion Batteries

2.1 Charging a Battery

In the case of rechargeable batteries, the battery must be charged in advance. When the battery starts to charge, electrons first move from the positive electrode (cathode) to the negative electrode (anode), and then ions move to the negative electrode. The important thing to note is that ions and electrons move along different paths. Ions move along paths inside solids and liquids, while electrons move along metal wires. This means that we can charge the battery by moving electrons with a charger from outside the battery. The electrochemical reaction occurs inside the battery. In the case of discharge, we can extract the electrons – and therefore the electrical power – and use them outside the battery (Figure 2.1).

2.2 The State of Stored Electricity and Discharge

What happens if you disconnect the charger after sufficient electrons and ions have moved from the cathode to the anode? The electrons will not be able to move, and the ions that have moved will not be able to

RSC Foundations No. 4
Li-ion Batteries
By Hitoshi Nakamura
© Hitoshi Nakamura 2025
Published by the Royal Society of Chemistry, www.rsc.org

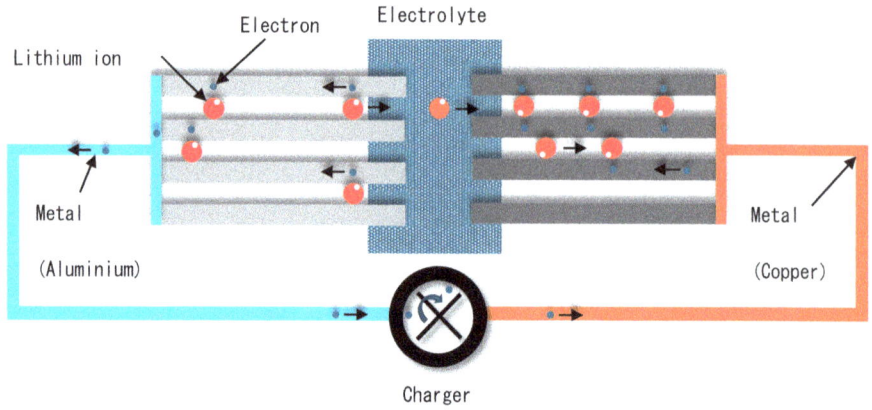

Figure 2.1   The charger moves electrons from the cathode to the anode. Not only do the electrons move, but also their energy increases as they move. The lithium ions that pair up with the electrons also follow the moving electrons and move together.

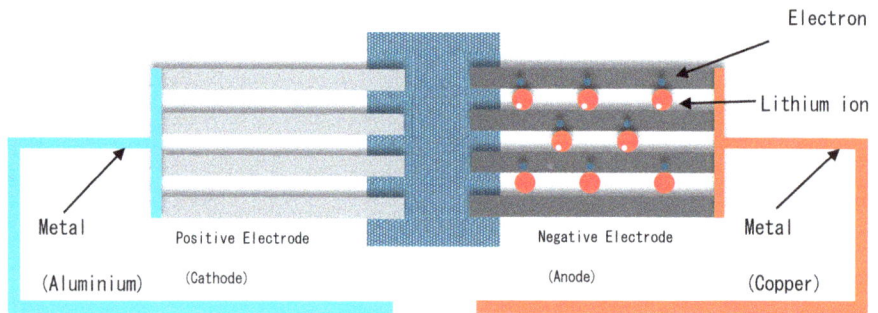

Figure 2.2   If the charger is removed (*i.e.* switched off) when the electrons and lithium ions have finished moving to the negative electrode (anode), the electrons and lithium ions stop directly at the anode. In other words, electricity is accumulated.

move either, because they are attracted to the electrons. This means that electrical energy has been stored (Figure 2.2). When the motor is placed between the cathode and anode and then switched on again, the electrons first move in the opposite direction, that is, toward the positive electrode. At this point, the electrons pass through the coil of the motor, and the motor turns. When the electrons return, the ions also follow the electrons and move toward the positive electrode (Figure 2.3).

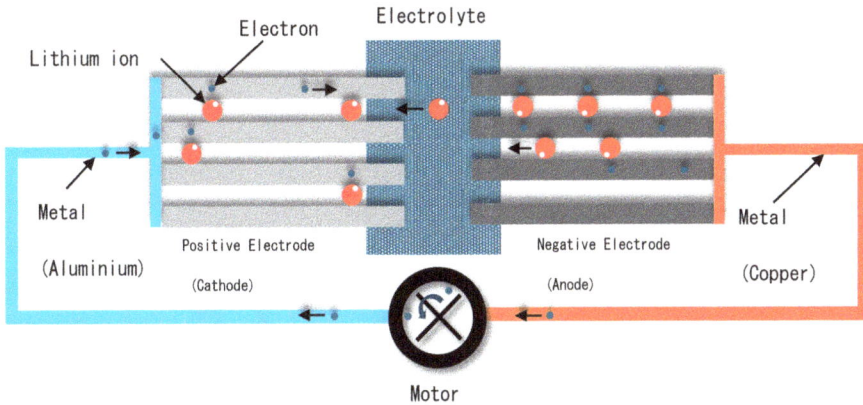

Figure 2.3   Next, when a load such as a motor is attached between the positive and negative electrodes, electrons move from the negative to the positive electrode. At this time, energy is lost by turning the motor, *etc.*

2.3   About the Liquids Through Which Ions Pass

The technical term for a liquid through which ions pass is "electrolyte". For example, dilute sulfuric acid is used in lead-acid batteries, which are widely used to start automobiles. Sulfuric acid is dissolved in water, which allows hydrogen ions to move through the electrolyte. By the way, when two electrodes are inserted into water with salt and a voltage of 1.2 V or higher is applied, bubbles are emitted from the electrodes, and decomposition begins. At the positive electrode, water is oxidized and decomposed to produce oxygen, and at the negative electrode, water is reduced to produce hydrogen. When building a battery with a water electrolyte, the potential of the cathode and anode must be kept between the oxidation and reduction potentials of water. If the potential of the positive or negative electrode is outside this range of voltages, the electrolyte begins to decompose. The range between the oxidation and reduction potentials of the electrolyte is called the "potential window". The potential window of the electrolyte in a lithium-ion battery is 4 V or higher, covering the potential of the positive and negative electrodes of the battery. Lithium-ion batteries use a liquid made of an organic substance (a compound containing carbon) as a solvent, to prevent decomposition even at high voltages.

Lithium ions are dissolved in this liquid and can move. It may seem strange to say that ions dissolve in organic matter, and this theory will be explained in Chapter 3.

2.4 Summary of Lithium-ion Battery Potential Correlations

Up to this point, I have been discussing the movement of electrons and lithium ions. Now I will explain how the voltage of the battery changes as the electrons and ions move. When a negative electrode made of carbon and a positive electrode made of a metal oxide containing lithium are immersed in an appropriate electrolyte, the voltage is almost zero. When voltage is applied to both electrodes, as explained earlier (see Figure 2.1), electrons move from the positive electrode to the negative electrode, and lithium ions move to the negative electrode in conjunction with this. The potential of the oxide from which the lithium ions have been removed changes, and the potential shifts in the positive direction. The potential of the carbon in the negative electrode shifts in the negative direction as it absorbs lithium ions. Eventually, the difference in potential between the positive and negative electrodes reaches 4 V or more. This process is called initialization. In the case of lead-acid batteries, the voltage also changes due to changes in the energy inside the battery. Similarly, in the case of lithium-ion batteries, the voltage fluctuates considerably. Figure 2.4 shows the charging process. All potentials are converted from the redox potential of lithium as 0 V. Lithium ions become the original lithium metal when the potential drops below the "redox potential" and lithium loses electrons and becomes an ion when the potential rises above the "redox potential". The positive and negative electrodes (materials) are immersed in an electrolyte. The standard hydrogen potential for lithium is −3.045 V. In terms of the science, it is more correct to use hydrogen as the standard for potential, but in the case of lithium-ion batteries, it is more convenient to use the potential of lithium as zero, so from now on, I will use the potential of lithium as zero. The potentials of the positive and negative electrodes are approximately 4.2 V and 0.1 V, respectively. The potential difference between the positive and negative electrodes is the cell voltage. The oxidation limit of the electrolyte is about 4.5 V, and the reduction limit is about 0.5 V. Since the reduction limit of the electrolyte is 0.5 V, the electrolyte decomposes at potentials below 0.5 V.

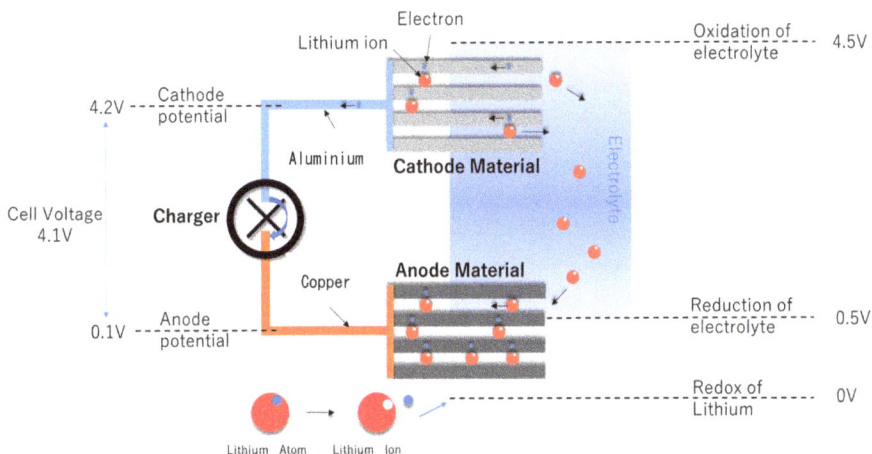

Figure 2.4   A summary of the explanation of the components of a lithium-ion battery, with relevant voltages.

Since the potential of the negative electrode is 0.1 V, the electrolyte is usually not retained. Lithium-ion batteries have been improved by prolonging the life of the electrolyte, but this is discussed separately in Chapter 3. The diagram shown in Figure 2.4 is very important and should be studied carefully.

2.5   Summary of Operating Principles

The operating process of a lithium-ion battery can be summarized as follows. When a lithium-ion battery is charging, an external voltage is applied, and electrons begin to move from the positive to the negative electrode. Lithium ions paired with the electrons follow the electrons out of the cathode and begin to move through the organic electrolyte to the anode. When the voltage reaches a level at which the anode can absorb the lithium ions, the ions are sucked into the gaps within the anode, which is structured like the pages of a book. This phenomenon is sometimes called "intercalation". When enough ions are sucked into the anode, the charging switch is turned off and electrons can no longer move, and thus energy is stored. When a motor or other device is placed between the cathode and anode of the battery and switched on, electrons flow from the anode to the cathode, and the motor rotates. Ions follow the movement of electrons and move from the anode to the cathode in the same way (Figure 2.5 and Video 2.1, https://www.youtube.com/watch?v=lSCLLtlwKmA).

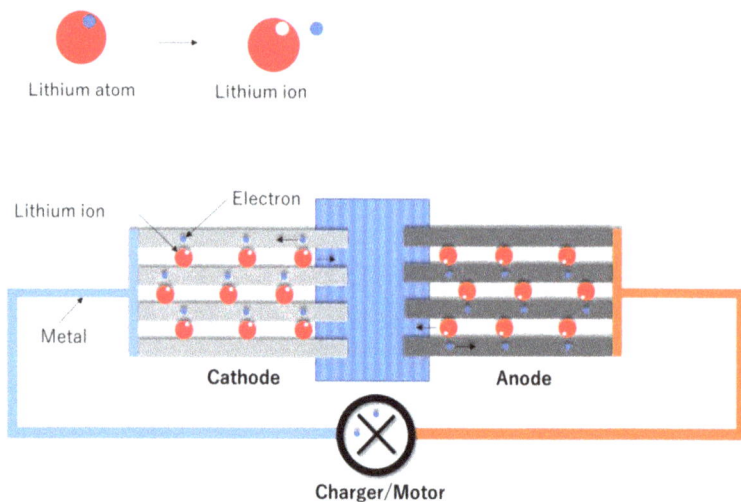

Figure 2.5 Summary of the operation of a lithium-ion battery.

2.6 Are Lithium Ions and Electrons Bound in the Electrode?

During charging, electrons move from the cathode, followed by lithium ions. During discharging, electrons move to the cathode, followed by lithium ions. This is how lithium-ion batteries work, but can the lithium ions that move after the electrons meet the electrons safely? Many studies have been conducted on this issue.[1,2,4] What state are the lithium ions in that are absorbed into the cathode during charging? The cathode material in a lithium-ion battery has a structure in which lithium is sandwiched between layers of metal oxide. To begin with, the metal oxide is a strongly acidic substance, and the strongly alkaline lithium neutralizes it. Looking at this in a little more detail, the metal oxide, which is negatively charged compared to the lithium ion, attracts the positively charged lithium ion and arranges it neatly in the gaps inside, like the pages of a book. If the bond between the lithium ions and the metal oxide is too strong at this point, the metal oxide and lithium ions will not separate even if a voltage is applied, so the ions will not move, and the battery cannot be used.

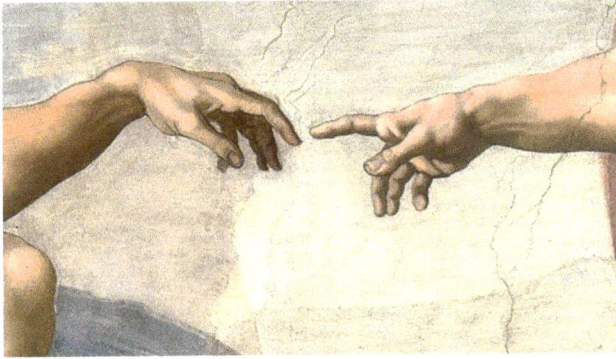

Figure 2.6 An exquisite relationship that seems to touch but cannot touch. Lithium ions never meet electrons in a battery. When they do, it is not a good thing. Image from Michelangelo's Creation of Adam (Public Domain).

Also, it is difficult to say whether the lithium ions absorbed into the negative electrode will be able to successfully meet up with electrons, but it is thought that they will not be able to meet up with electrons and will remain in ion form. At first glance, this seems like an unfortunate occurrence.

However, what if the lithium ions and the electrons in the negative electrode were to successfully meet their goal and return to their original metallic state? In this situation, the metallic lithium would grow and penetrate the separator membrane, and, in the worst-case scenario, the positive and negative electrodes would short circuit, leading to overheating, smoke, and even fire. Unfortunately, the only way to prevent this is to keep the electrons and lithium ions almost completely separate (Figure 2.6). But you might be asking, "Where do the electrons come from if the lithium ions are still ions?" During charging, the ions move around in the positive electrode, causing the surrounding metal to release electrons instead. In the negative electrode, the electrons and ions are attracted to each other without ever meeting.[1,3,5]

Lithium-ion batteries have a higher output than other batteries. One reason for this is that the lithium ions are only weakly attracted to their hosts on both the positive and negative electrodes. The name "lithium-ion battery" was coined a few years after Dr Yoshino, Dr Goodenough and Dr Whittingham's invention, when Sony launched its first lithium-ion battery. I don't know if the developers at the time were aware of the situation, but their naming sense is quite profound.

References

1. Y. Maniwa, K. Kume, H. Suematsu and S.-i. Tanuma, High resolution ^{13}C NMR in K-graphite intercalation compounds— c-axis charge distribution, *J. Phys. Soc. Jpn.*, 1985, **54**, 666.
2. C. Liu, Z. G. Neale and G. Cao, Understanding electrochemical potentials of cathode materials in rechargeable batteries, *Mater. Today*, 2016, **19**(2), 109–123.
3. A. Manthiram, A reflection on lithium-ion battery cathode chemistry, *Nat. Commun.*, 2020, **11**, 1550.
4. H. Moriwake, A. Kuwabara, C. A. Fisher and Y. Ikuhara, Why is sodium-intercalated graphite unstable?, *RSC Adv.*, 2017, 7, 36550.
5. M. Saint Jean, C. Fretigny and M.-F. Quinton, 13C NMR or- bital shift calculation in graphite intercalation compounds, *Mol. Cryst. Liq. Cryst. Sci. Technol., Sect. A*, 1994, **245**, 123.

3 Components and Materials That Make up a Lithium-ion Battery

There are four major materials that make up a battery, and this allows us to establish a basic system of lithium-ion batteries. In addition to the cathode, anode, and electrolyte described in previous chapters, there is the separator. The separator is an insulating film that prevents the cathode and anode from touching each other directly. These four major components are discussed in this chapter. Lithium-ion battery systems are averse to moisture and require a case to isolate them from the general environment, which is discussed in Chapter 5.

3.1 Positive Electrode Material

To use a lithium-ion battery, it must first be charged. When a lithium-ion battery is charged, electrons move from the positive electrode to the negative electrode, and lithium ions move at the same time, so the positive electrode must contain lithium ions in advance. The pre-contained lithium is the very characteristic of the cathode for a lithium-ion battery. The interior of a lithium-ion battery is structured like a stack of paper, as in a book, and lithium is stored in the gaps between the layers.[1] (Some batteries may use spaces that look like stacked blocks or tunnels, but I will explain that later.) Unlike paper,

RSC Foundations No. 4
Li-ion Batteries
By Hitoshi Nakamura
© Hitoshi Nakamura 2025
Published by the Royal Society of Chemistry, www.rsc.org

The TEM Image of $LiCoO_2$ The TEM Image of Graphite

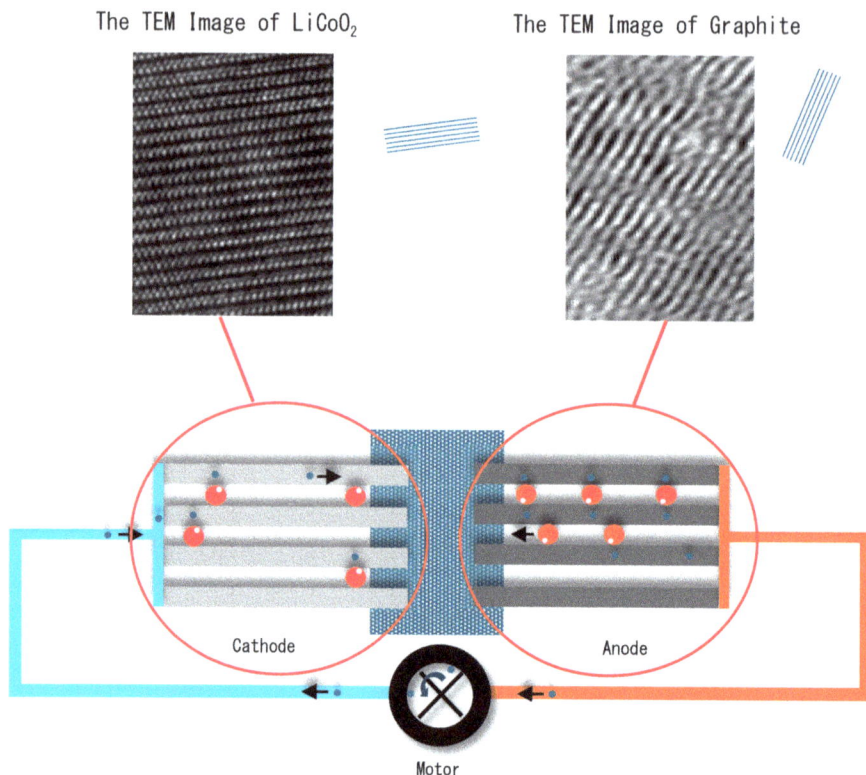

Figure 3.1 Both the positive and negative electrodes have a structure like the pages of a book, and lithium ions are stored in the gaps between the pages. Electrons are so small that they exist in the structures that make up the book. Structures that create gaps where ions exist include block-like and tunnel-like structures, which will be explained in Chapter 4.

the thin walls are made of a material called a semiconductor,[2] which allows electrons to pass through (Figure 3.1). The potential at which the positive electrode adsorbs and desorbs lithium differs depending on the material used. Since higher voltage is better, the potential at which the cathode absorbs and de-absorbs lithium ions should be as high as possible. The voltage at which lithium becomes an ion (loses electrons and is oxidized) and the voltage at which lithium ions gain electrons and return to atoms (metal) is called the redox potential of lithium. Based on the redox potential of metallic lithium, the positive electrode is generally around 4 V to 4.5 V. In other words, if metallic lithium is used for the negative electrode, a 4–4.5 V battery can be constructed. The potential of the cathode is determined by the type of metal that makes up the oxide. Metal oxides are so-called ceramics. The metals include cobalt, nickel, manganese, and iron. As discussed below, most cathode materials are produced by firing, as in pottery.[3]

3.1.1 Relationship Between the Number of Electrons and the Capacity in 'mA h'

The capacity of a cathode is a measure of how many lithium ions can be absorbed by the cathode material. The unit of measurement is mA h g^{-1}, which indicates how many mA h of ions can be absorbed per unit gram (or discharged in discharge). Now, we will calculate how many lithium ions are equivalent to how many 'A h', as shown in Box 3.1.

Since lithium emits one electron when it becomes an ion, it can be seen that 2.2×10^{19} ions must be transferred to produce a current of 1 mA. The theoretical capacity is an indicator of how many mA h of electricity can be stored by the cathode at its maximum. Using lithium cobaltate, $LiCoO_2$, a typical cathode material used in lithium-ion batteries, as an example, the theoretical capacity can be calculated as shown in Box 3.2.

This calculation shows that the theoretical capacity of lithium cobaltate is about 274 mA h g^{-1}. The conversion is complicated because it involves converting from moles of electrons to mA h, but, for now, just remember that one mole of electrons (one mole of lithium ions) is about 26.8 A h, and then a simple proportional calculation can be performed.

Box 3.1 Calculating how many electrons move when 1 A h of current flows.

According to Faraday's law, one mole of ions (= electrons) is equivalent to 96 500 coulombs (= A × seconds).
 This is converted to 26.81 coulombs (= A × hour).
 1 mole (of electrons) is equivalent to 6×10^{23} electrons.
 If 1 A h of current has been flowing, then 2.24×10^{22} electrons were transferred.

Box 3.2 Calculating the theoretical capacity of lithium cobaltate.

The molecular weight of $LiCoO_2$ is 97.8 g mol^{-1}.
 One mole of $LiCoO_2$ contains 1 mole of lithium ions.
 One mole of electrons is 96 500 coulombs (= A × seconds).
 This is equivalent to 26.8 A h [96 500 coulombs (= A × seconds)/3600 seconds = 26.8056 A h per coulomb].
 Which can be converted to 26 805.6 mA h.
 Dividing the electric quantity (mA h) by the molecular weight gives 274.1 mA h g^{-1}.

3.1.2 About Oxidation and Reduction

Several definitions of acids and alkalis appear in school textbooks. As our understanding of acids and alkalis progresses, we move from the Brønsted–Lowry model, based on ions, to the more essential Lewis understanding, based on the gain or loss of electrons. Since a battery is a device that deals with electrons, this book will be based on the gain or loss of electrons. In other words, loss of electrons is considered to be oxidation, while gain of electrons is considered to be reduction.

3.2 Anode Electrode Material

3.2.1 Carbon Anode Material

Carbon has been used as the anode of lithium-ion batteries. The higher the crystallinity of carbon, the lower the potential at which it can absorb lithium and the higher the potential[1] difference with the positive electrode (*i.e.* the battery voltage). Graphite, a highly crystalline carbon material, has the lowest potential to absorb lithium, at about 0.1 V based on the redox potential of lithium. Thus, combined with the 4.2 V cathode described earlier, a battery with a voltage of about 4.1 V can be configured.

Graphite has a structure resembling stacked sheets of graphene, and graphene consists of carbon hexagonal rings arranged in an infinitely extended structure. (Figure 3.2).

It should be noted first that the classification of carbon types is somewhat complicated, and some definitions are not clear. According to the literature,[4] graphite is defined as having a film-to-film distance (inter-layer distance) of less than about 3.335 angstroms (or 0.335 nm).

The radius of the lithium ion is 0.59–0.92 angstroms, so it would seem to fit between the layers of carbon without any problem, but the space that the lithium ion can enter due to the electrons around the graphene (these electrons are not involved in the interaction

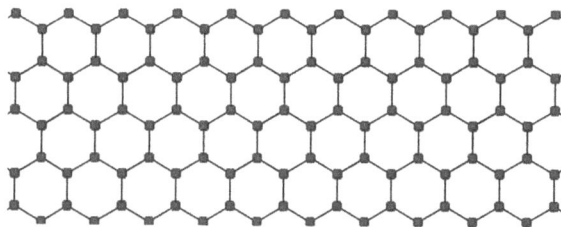

Figure 3.2 Graphite is composed of layers of horizontally arranged benzene ring films (called graphene).

with the lithium ion but are responsible for linking carbon atoms together) is the same as above and it is believed to be narrower than its width. In fact, when lithium ions enter the interlayer of graphite, the distance between the layers widens and the graphite expands by up to 10% in the vertical direction.[7,8] In addition, when viewed from the top surface of the graphite, the lithium ions stay in the center of the hexagons (Figure 3.3) because there are electron films around the hexagons that interfere with the lithium ions. The molecular formula of lithium ion-adsorbed carbon (graphite) is expressed as C_6Li, with the smallest unit being six carbons and one lithium ion.

Since the diameter of lithium ions is slightly larger than that of the graphite interlayer (as mentioned above, the interlayer distance including the electronic disincentives), the lithium ions cannot enter the interlayer as it is. However, lithium ions attract electrons, and this force pulls lithium ions into the interlayer (Figure 3.4).

As electrons are charged, the negative potential decreases, and the force that tries to pry open the interlayer overcomes the force that blocks lithium ions, and lithium ions enter the interlayer. Therefore,

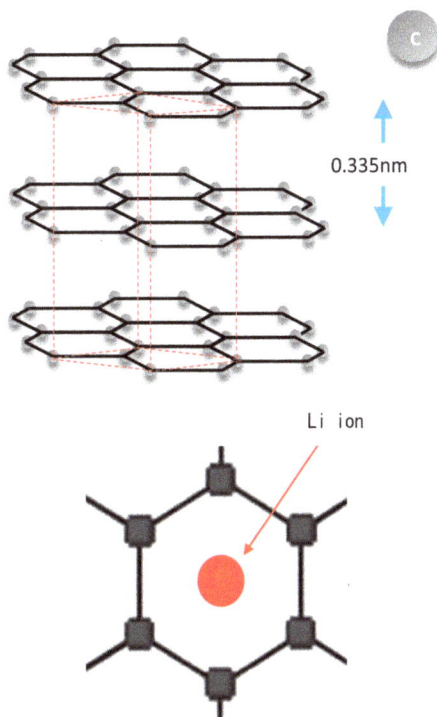

0.335nm

Li ion

Figure 3.3 The lithium ion cannot exist anywhere on the graphene plate, but sits right in the middle of the six-membered ring, blocked by the electrons of the carbon that make up graphene.[5,6]

the potential for lithium ions to enter depends on the interlayer distance of the graphite, although it is subtle. So, the amount of lithium ions that can be absorbed, the potential at which the absorption begins, or the speed at which the ions diffuse into graphite varies slightly depending on the crystallinity of the graphite.[4]

The image in Figure 3.4 is only narrative, but the concentration of electrons transferred to the negative electrode by the charger increases, and the force is to bring in lithium ions to compensate for this imbalance.[4]

Based on this information, the theoretical capacity of graphite is calculated to be a little more than 372 mA h g^{-1} from the calculation shown in Box 3.3.

Graphite can be broadly classified into natural graphite and artificial graphite. Natural graphite is synthesized deep underground under high pressure and temperature, and has higher crystallinity than artificial graphite. Therefore, the capacity obtained from natural graphite is greater. Artificial graphite is obtained by subjecting a carbon source, such as coal coke, to high temperatures of 2000 °C to 3000 °C to increase its crystallinity. However, the crystallinity is not as high as that of natural graphite and so the capacity is slightly lower. On the other hand, the distance between the layers of carbon is slightly greater in artificial graphite than in natural graphite, which facilitates the movement of lithium ions, making it suitable for applications that require high power output, such as battery electric vehicles (BEVs). In general, natural graphite is less expensive, and natural graphite and artificial graphite are often blended for consideration of the required cost and performance.

Figure 3.4   Image of electrons pulling in lithium ions.

Box 3.3   Calculation for determining the theoretical capacity of graphite.

The molecular weight of C is 12 g mol^{-1}.
   Because six carbon atoms can contain one lithium ion, this is equivalent to 0.17 mole of lithium ions in each molecule.
   One mole of electrons is 96 500 coulombs (= A s).
   This is equivalent to 26.8 A h.
   Which is converted to 26 805.6 mA h.
   Dividing the electric quantity by the molecular weight gives 372.3 mA h g^{-1}.

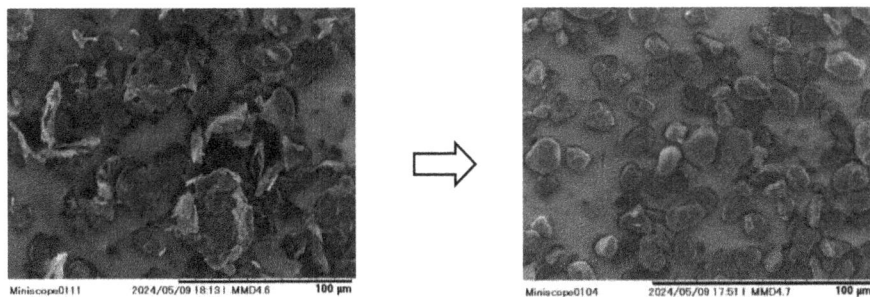

Figure 3.5 An example of spheroidized graphite material made from leaf-shaped graphite.

The shape of natural graphite is leaf-like, which is thin and flat. When lithium-ion batteries are to be installed in vehicles, the energy density per volume of the battery is a key factor, as well as the weight. When trying to pack a large number of particles into a fixed space, it is advantageous to use a spherical shape. For this reason, graphite is also processed into a spherical shape (Figure 3.5).

In order to increase the density of the anode and reduce side reactions, graphite is spheroidized and coated with amorphous carbon on the surface. There is an interesting way to mechanically round the original leaf-shaped graphite as well, but I will not go into that here. A few Japanese manufacturers launched this product in the 1990s[9] and this series of graphite-related processing technologies is still in use around the world.

3.2.2 Silicon Anode Material

Silicon can achieve a capacity of over 4200 mA h g^{-1}, up to ten times that of graphite.[10] The potential at which it can absorb lithium ions is also close to that of graphite. On the other hand, silicon is often used mixed with graphite because of its larger volume expansion when lithium ions are sucked in, compared with graphite. Currently, the mixing ratio of silicon used in batteries is about 20%.[11,12] This mixing ratio is expected to increase with technological innovations in the future.

3.3 Energy Density of Lithium-ion Batteries

Since energy is the multiplication of voltage and electric capacity (eqn (3.1)), batteries with the highest possible voltage and capacity are preferred.

$$E = V \times A h \left(W h \right) \tag{3.1}$$

In designing batteries, it is difficult to achieve both energy density and output density, and there is always a trade-off between the two. Some lithium-ion batteries for cell phones, which emphasize energy density, now exceed 300–400 W h kg^{-1}. When looking at battery specifications, these figures can be used as a reference to understand the quality of the battery. In the past, materials used in Li-ion batteries have been researched and optimized for portable applications, and so I do not expect any drastic changes in the future. However, dramatic improvements are being made, such as by changing the negative electrode to metallic lithium; but a battery with a metallic negative electrode instead of a carbon negative electrode cannot be classified as a lithium-ion battery, from an historical perspective. Thus, I will not go into detail here.

3.4 About Separators

If the cathode and anode are in direct contact with each other, electrons go directly between the electrodes instead of through the wires, and, naturally, electricity cannot be extracted and used outside (*i.e.* this is not viable as a battery). Therefore, the cathode and anode must be physically separated. Lithium ions move in the electrolyte, so the cathode and anode must be connected by the electrolyte. In practice, a sheet made of electrically insulating plastic is sandwiched between the cathode and anode to insulate them electronically. Although the electrons are insulated, the ions must be able to pass through, so the sheet has small holes. In other words, it is a membrane that does not allow electrons to pass, but does allow ions to pass (Figure 3.6).

There are two methods of making separators: a dry method and a wet method. In the dry method, a thin sheet of plastic is pulled from side to side to create a tear of appropriate size (this is called stretching). In the wet method, the sheet is mixed with a finely ground solvent-soluble material in advance, and then the sheet is soaked in the solvent to dissolve only the soluble portions to create holes. A combination of the dry and wet methods is also used.

When the positive and negative electrodes of a lithium-ion battery, which store energy, come into contact with each other, a large amount of electrons is transferred and a large amount of heat is generated (this is called an internal short circuit). Half of all fires involving batteries

Figure 3.6 High-resolution scanning electron microscopy (SEM) image of the separator surface. Photo courtesy of Dr Jyotishman Pathak.

are caused by internal short circuits. For this reason, the separator plays a very important and somewhat contradictory role in insulating the positive and negative electrodes electronically while still allowing a large amount of lithium ions to pass through. The separator is made of polypropylene (and polyethylene) film, which is insoluble in the electrolyte. Polypropylene melts at high temperatures, and there is a risk of an internal short circuit when the cathode and anode come into contact with each other.

An important role of the separator is its shutdown function. In particular, when a lithium-ion battery heats up for some reason during charging (the non-safe mode of lithium-ion batteries is explained later in Chapter 8), the separator melts and blocks the passage of ions, thereby blocking the passage of lithium ions and inhibiting their reaction. In the photograph (Figure 3.7), it can be seen that as the heat rises, the openings in the separator unravel and become clogged. The shutdown function can also be achieved with polypropylene material, but to achieve a sharper shutdown function, a three-layer structure with polyethylene sandwiched in the center is used, so that the polyethylene melts first and blocks the lithium ion migration (Figure 3.7). Early patents for separators with a shutdown function were registered by Dr Yoshino's group (he was jointly awarded the Nobel Prize).[13]

Recently, a thin coating of fine, heat-resistant ceramic powder has been used on the surface of the sheet, helping to improve the safety of lithium-ion batteries (Figure 3.8).

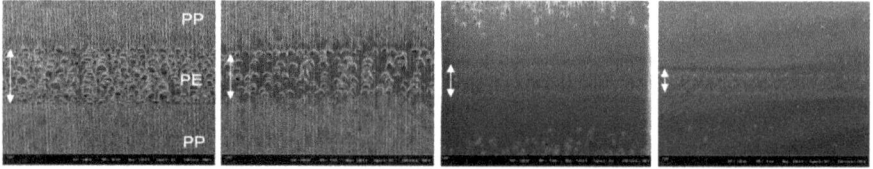

Figure 3.7 The three-layered separator is melted first from the polyethylene layer by heat at 130 °C, and the voids disappear.[36] Reproduced from ref. 36 with permission from ITES Corporation.

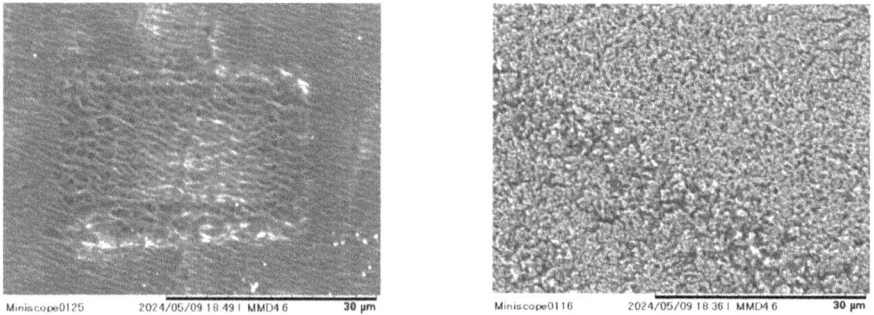

Figure 3.8 (Left) Surface of a separator formed by the dry process (stretching method). The shish-kebab structure is seen. (Right) The surface of a separator with a thin coating of alumina powder is shown. This improves the heat resistance of the separator and prevents shrinkage, dramatically improving safety.

3.5 About the Electrolyte

As mentioned in the previous section, the electrolyte is responsible for passing lithium ions. When charging, electrons move first, and then lithium ions, following the electrons, reach the negative electrode. Since it is wasteful to wait for the lithium ions to reach the negative electrode, a certain amount of lithium ions are dissolved in the electrolyte in advance, and during charging the ions from the positive electrode push the ions in the electrolyte one after another, like dominoes, and they are sucked into the negative electrode.[14] This ensures efficient transfer of lithium ions. The output (power) of a battery is how many electrons or ions can be moved in a given period of time. Since electrons in a battery move much faster than lithium ions, the power of a battery depends on how fast lithium ions can move. The electrolyte plays a role in transporting lithium ions and is thus closely related to the power of the battery. The positive electrode of a lithium-ion battery has a high potential range of 4 V or more, relative to the redox potential of lithium, while the negative electrode is near 0 V. The electrolyte must transport lithium ions across the positive and negative electrodes, so it must be stable over a wide range of potentials. The electrolyte is made of a material that dissolves in a solvent to form ions (called a salt because it is a combination of lithium and PF_6^- ions, analogous to the 'salt' of sodium and chlorine ions). The range of potentials that a solvent can withstand for oxidation and reduction is called the potential window. When water is mixed in the electrolyte, the operating voltage range of a lithium-ion battery is wider than the potential window of water (about 1.2 V), so water decomposes in the organic electrolyte and causes various problems. Avoiding water contamination is very important for stable operation of lithium-ion batteries. Lithium ions are paired with electrons in the electrode active material, but in the electrolyte they are paired with negative ions. PF_6^- ions are mainly used as negative ions. For the lithium ion to move, the counter negative ion must also move, in the opposite direction (if the lithium ion moves toward the negative electrode when charging, the counter negative ion moves toward the positive electrode). The speed at which negative ions move also greatly affects the speed at which lithium ions move.[15] The solvent of the electrolyte is called a polar solvent. Polar solvents have a positive and a negative part within one solvent molecule. For example, substances with the chemical formula RCOOR, as shown in Figure 3.9, are called esters. Esters have a double bond between oxygen and carbon. The oxygen attracts electrons from carbon, and the density of electrons is biased such that the oxygen at the end of the double bond is negatively charged ($\delta-$) and the carbon is positively charged ($\delta+$).[16]

This polar part attracts lithium ions and can dissolve more lithium ions. If an ester has a ring in it, a cyclic ester, the negative tendency of the oxygen at the end of the double bond becomes even stronger, so it can attract and dissolve a large number of lithium ions. However, the ease of attracting lithium ions also means that the ease of inhibiting the movement of lithium ions, and the movement of lithium ions in the cyclic ester, is poor. Therefore, linear esters, which allow lithium

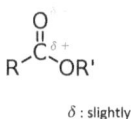

$$R-\underset{\underset{OR'}{|}}{\overset{\overset{O}{\|}}{C}}_{\delta+}$$

δ : slightly

Figure 3.9   The minute polarization in an ester molecule.

ions to move easily, are blended in appropriate amounts and used. In addition, the movement of lithium ions in the solvent slows down at low temperatures, so if the battery is to be used at low temperatures, the proportion of linear esters should be increased. However, linear esters tend to be reduced and decompose more strongly than cyclic esters, so there is a concern about battery deterioration. The know-how for blending solvents according to the proposed battery use is a matter of skill for electrolyte manufacturers.

3.6   About Solid Electrolytes

In research on solid electrolytes, it has often been pointed out that, although the ionic diffusion resistance of the solid electrolyte itself is sufficiently low, the interface resistance can be a problem, and, in some cases, the apparent characteristics cannot be obtained. Many glass-like sulfide-based solid electrolytes are more flexible than crystalline oxide-based solid electrolytes. This flexibility is important for interfacial formation and promotes lithium ion transfer by adhering to the surface of the positive and negative electrode active materials. Recently, attempts have been made to make oxide-based solid electrolytes that are softer by adjusting the sintering temperature and by other means. The disadvantage of sulfide-based electrolytes is that they are vulnerable to moisture and are prone to deterioration in their characteristics and the generation of hydrogen sulfide gas. Solid oxide electrolytes are easy to handle and very safe, but they are hard and have poor adhesion with active materials. For this reason, attempts to compensate for the interface by injecting a small amount of electrolyte into the solid oxide electrolyte and using the electrolyte have been relatively successful. One of the major advantages of all-solid-state batteries is that they are less likely to catch fire, even if an accident occurs, and, even in hybrid models, the amount of electrolyte that can be used as fuel in the event of a fire is small, so the damage caused by fires can be minimized.

3.7   About the Current Collector

The cathode or anode active material is made of a fine powder. To compensate for the conductivity of the powder, it is mixed with a small amount of conductive auxiliary material. Next, a small amount of adhesive is used to bond the particles together. This is called a composite material. A thin metal that transfers electrons is bonded to the composite (Figure 3.10). This thin metal is called the current collector. The current collector must allow electrons to pass through easily and must be stable in the potential between the positive and negative electrodes. Historically, aluminium has been used for the positive electrode and copper for the negative electrode. The redox potential of aluminium is around 1.3 V relative to the redox potential of lithium. In other words, if a voltage of 1.3 V or more is applied with respect to the potential of lithium, the aluminium ions will dissolve. If the potential is raised to nearly 4 V with respect to lithium, aluminium should dissolve without a moment's hesitation. However, aluminium has the property of forming an oxide film on its surface to protect the interior, so, in practice, it can be used without any problems. Although it can be used on the reduction side, aluminium and lithium react at around 1.23 V relative to the redox potential of lithium to form an alloy, making it unusable as a negative electrode current collector for a lithium-ion battery. Asahi Kasei, where Dr Yoshino—one of the inventors of the lithium-ion battery—was employed, held this fundamental patent.[17] A major barrier of this patent was the use of aluminum as the current collector for the cathode. On the other hand, the potential of copper is around 3.03 V, which is stable enough for the negative electrode to work. For this reason, copper is used as a negative electrode

Figure 3.10 Cathode and anode of a lithium-ion battery. A thin layer of active compound is applied on the aluminium and copper collecting electrodes, respectively.

collector. However, if the battery is over-discharged and the negative electrode exceeds 3 V, copper may melt and react with the electrolyte, releasing hydrogen gas and causing the battery to swell.

3.8 Additives and the Anode Interface

I explained above that the solvent is resistant to oxidation and reduction. In fact, there is no solvent that can withstand reductive decomposition (reaction in which electrons are forced to decompose) near the potential (0.1–0.2 V) at which lithium ions are inserted into and extracted from the graphite negative electrode, and most solvents decompose on the negative electrode surface when the battery is charged. However, the decomposed solvent covers the anode surface as a kind of film (called the solid–electrolyte interface (SEI)). This film protects the cathode surface, so that after the initial decomposition of a certain amount of electrolyte, the reaction becomes less intense. However, the film that is made in the natural process is weak, which can lead to problems, such as shortening of the battery life. Therefore, it was thought that it would be a good idea to mix a few percent of additive into the solvent that would decompose by reduction at an earlier stage than the electrolyte so as to form a stronger film, and this was found to have a significant effect.[19–21] A typical example of an early additive is vinylene carbonate (VC)[22] (Figure 3.11), which decomposes reductively before the solvent decomposes, forming a film. The film inhibits decomposition of the solvent and suppresses the adverse effects associated with solvent decomposition (*e.g.* capacity degradation, power loss, and battery swelling due to gas generation).

In my explanation of the operating principles of lithium-ion batteries, I wrote that the use of graphite with high crystallinity allows the use of low potentials, thus increasing the battery voltage. At first, it was difficult to use graphite because the solvent decomposes too rapidly. This is generally considered to be due to the catalytic effects of the graphene edge surface. The use of additives has made it possible

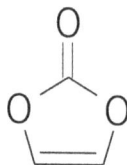

Figure 3.11 Molecular formula of vinylene carbonate.

to use graphite stably, with the energy density of lithium-ion batteries being dramatically improved by a factor of about 2, compared to the initial invention, thanks to additives that allow graphite to be used.

In addition to using an additive to protect the surface of the anode graphite, by applying a pre-coating, coating the surface of the anode graphite with some kind of film beforehand was also proposed. Graphite with high crystallinity decomposes the electrolyte, whereas low-crystallinity carbon is less reactive with the electrolyte, so a coating of low-crystallinity carbon on the graphite would be sufficient. Naturally, combining the surface-treated graphite material with an additive will give better results. One of the groups that came up with the electrolyte additive and graphite surface treatments was that of Professor (now Professor Emeritus) Yoshio of the Faculty of Science and Technology at Saga University, and they achieved great results.[23]

3.9 Solvation of Lithium Ions

Solvent molecules have polarity and are strongly attracted to lithium ions, and so can dissolve them. The solvent near the lithium ion is tightly bound and forms a large molecule. This state is called solvation.[24] The image of lithium moving by dragging several molecules of solvent with it is the conventional theory. The concept of solvation is valid in dilute solutions, but is not necessarily valid in concentrated salt systems. Given the output power response of lithium-ion batteries, the concept of lithium ions hopping between solvents has been proposed, and this understanding has become mainstream in recent years.[14,15] It is, however, worth remembering the concept that lithium ions behave as large molecules due to solvation (Figures 3.12 and 3.13).

The hopping model explains the fast ion migration seen in batteries, which cannot be explained by using the solvation state.[25]

3.10 Prospects for Cathode and Anode Materials

NEDO, an affiliate of Japan's Ministry of Economy, Trade and Industry (METI), has enthusiastically subsidized lithium-ion battery material research in the past, and the data provided in the report from 2013[29] provide a good indication of future improvements in lithium-ion battery performance (Figure 3.14). Although these results are more than a decade old, the situation has not changed significantly. To give a

Solvation Model

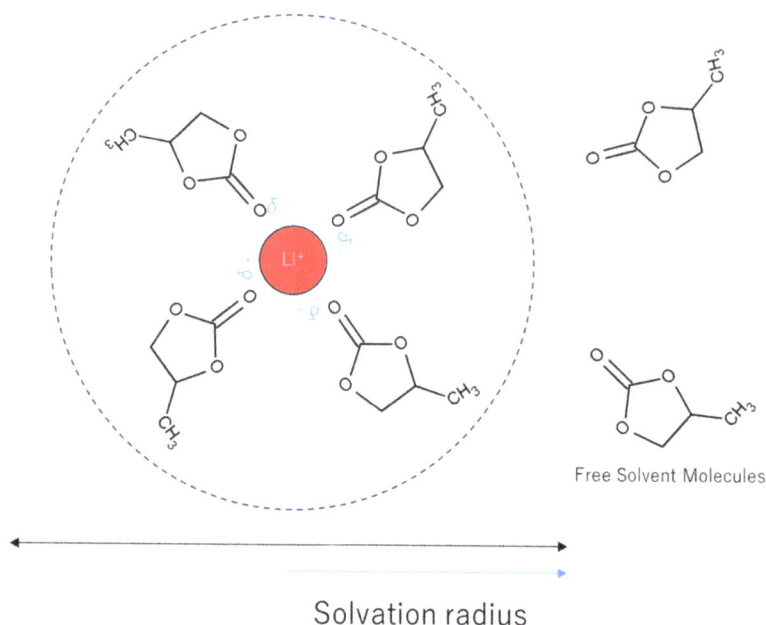

Free Solvent Molecules

Solvation radius

Figure 3.12 The solvation model showing the lithium ion moving and dragging some solvent with it.

quick overview, candidates for improving the negative electrode have emerged, and now graphite and silicon alloys are being used in a mixture. The ultimate anode is lithium metal, but there are various practical issues relating to its use. Compared to the anode, the development of the cathode has been slower and has stalled at around 200 mA h g^{-1}. Sulfur is considered to be the ultimate cathode material. However, the electrolyte used still poses many problems, and so sulfur has yet to be put to practical use. The all-solid-state battery is one approach to realize the Li–S battery. For the time being, it is likely to be used as an extension of a system that uses a liquid or solid electrolyte with cobalt, nickel, and manganese oxides, containing lithium for the positive electrode and graphite (plus a small amount of silicon or silicon alloy) for the negative electrode. Recently, with the rising cost of cobalt and nickel and other risks to resources, there has been a growing movement to use less expensive, high-capacity manganese solid-solution cathode materials and iron phosphate oxides. I will discuss these trends in more detail later in Chapter 8.

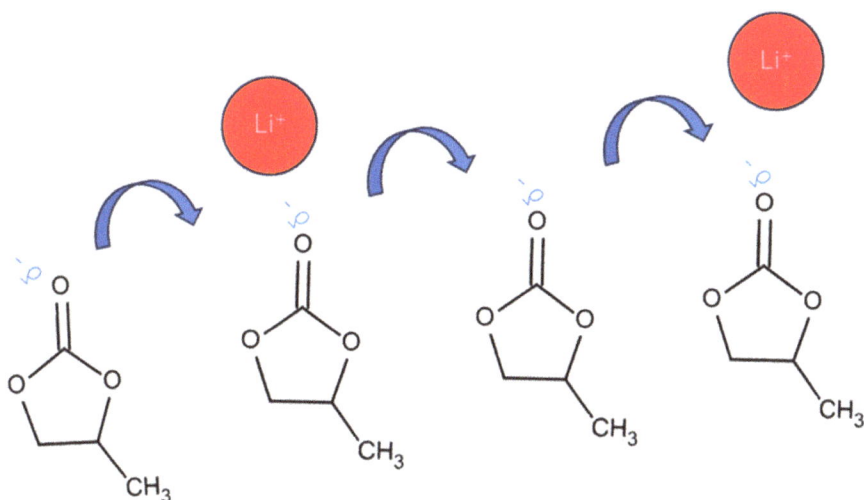

Hopping model

Figure 3.13 The hopping model in which the lithium ions do not move with the solvent, but actually hop along from one solvent molecule to the next.

3.11 A Bird's-eye View of Cathode Materials

As I explained earlier, the energy is the product of voltage and capacity and it is desirable for cathode materials to have an operating potential that is as high as possible, where lithium ions are sucked in and released. Typical cathode materials with high operating potentials and relatively high capacities are materials known as layered oxide systems, with a structure of layered oxides of transition metals, such as nickel, manganese, and cobalt, which overlap like the pages of a book, and with lithium ions sandwiched between the layers. This was developed by Dr J. B. Goodenough in the early days of lithium-ion battery invention. The earliest such layered oxide that was used in lithium-ion batteries was $LiCoO_2$.[26] Cobalt was used in semiconductors and other applications, and so it soared in price, and was then partially replaced by nickel, which has a higher potential than cobalt. However, the cathode partially replaced with nickel was more unstable than the pure cobalt-based cathode, so a material known as a ternary system, in which a portion of the cathode was replaced with

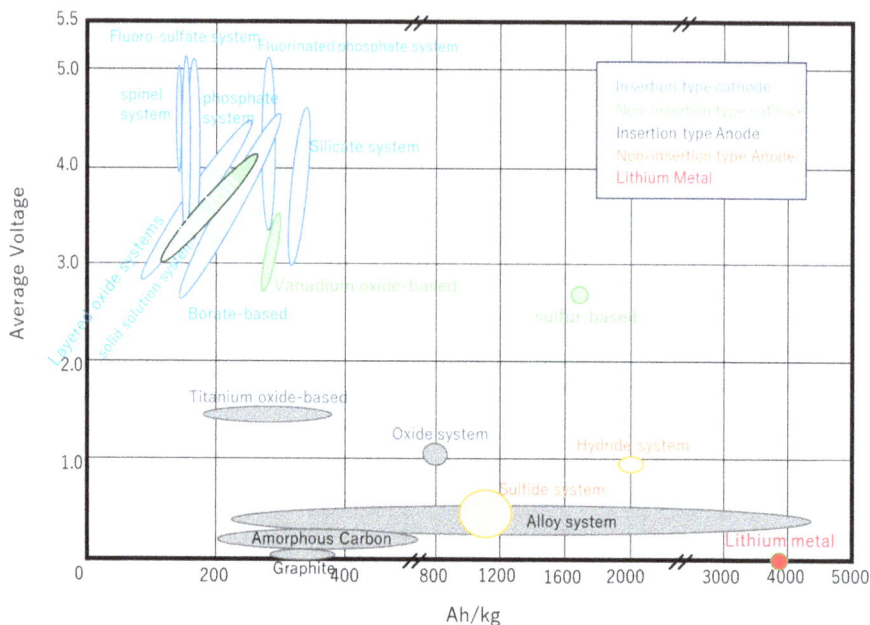

Figure 3.14 Comparison of active cathode and anode materials for use in Li-Ion batteries. Data from ref. 29. The capacity of the cathode candidates is relatively low compared to the capacity that can be achieved for the anode. This situation has not changed much since 2024.[29]

more stable manganese, became common. Materials stabilized by using aluminium instead of manganese were called nickel cobalt aluminium (NCA)-based materials. The capacity of these layered ternary materials is around 270 mA h g^{-1}, as calculated earlier. They are widely used in portable devices because of their high energy density, with an operating potential of 4 V or higher. In this book, the term "layered oxide system" or "ternary system" is used.

Manganese-based materials were also considered. Manganese-based materials can be used in layered systems, but they are unstable, and attempts were made to use spinel-shaped oxides, which are stable, for the cathode. Spinel is a way of classifying gemstones, and spinel crystals have an octahedral structure; imagine a structure of overlapping octahedra. Lithium ions move through the gaps between the octahedra. The molecular formula is $LiMn_2O_4$. The voltage of the spinel system is as high as 4 V, but the capacity is lower than that of the layered oxide system because of the two extra oxygen atoms. However, manganese is cheap, so it was used in early electric vehicles.

The iron phosphate system is cheaper than other systems because it uses iron. However, the voltage is as low as 3.3 V, so the energy density is one-half to two-thirds that of the layered oxide system. However, it is relatively safe because of its strong olivine structure. Since iron phosphate is safe and low cost, structures to increase its safety can be simplified, and it is now widely used in electric vehicles because it can be used efficiently despite its low energy content. Olivine is also a term derived from the classification of ores; when crystals overlap they form a structure that resembles a long, thin tube, like a straw. Lithium ions move through those tubes. If the voltage is low when iron is used, why not use manganese instead? However, lithium ions move slowly in manganese phosphate-based materials, so, in practice, hybrid materials of iron and manganese are used, which are called lithium iron manganese phosphate (LFMP)-based materials.

As for the cathode materials for lithium-ion batteries, layered oxide and iron phosphate were invented by Dr Goodenough. There are various theories about spinel-based materials, but Dr Yoshio, Professor Emeritus at Saga University, has made a significant contribution.[27,28]

3.12 Positive and Negative Electrode Utilization

We have calculated that the theoretical capacity, or maximum capacity, of cathode materials is approximately 274 mA h g^{-1} (see Box 3.2). However, the capacity of the cathode materials currently in use is around 200 mA h g^{-1}. I would like to explain what this means. Figure 3.15 shows the voltage–charge curve of a lithium-ion battery. As the voltage is increased, the change in voltage becomes steeper until the voltage is around 4.6 V, and the capacity is around 260 mA h g^{-1}. The voltage increase becomes steeper as the lithium ions in the cathode run out and the voltage change can no longer be neutralized. If the cathode material is of good quality, it can release lithium ions at a level close to the theoretical capacity. However, at present, only about 75% (200/274) can be used, resulting in a capacity of about 200 mA h g^{-1}. This percentage is called the utilization ratio. There are two reasons for the limited utilization rate. One is the resistance limit of the electrolyte to redox. The voltage of 4.6 V, which maximizes the capacity of the cathode, is almost at the limit of the voltage of the oxidation side of the electrolyte that can be withstood, and so cannot be maintained for very long. However, the user wants the cathode to be able to withstand at least 1000 charge–discharge cycles, so the voltage should be kept at around 4.2 V to allow for the oxidation resistance of the electrolyte, and, as a result, the capacity settles at around 200 mA h g^{-1}. The second reason is as follows. The cathode material of a lithium-ion battery is stable because lithium ions are contained in the gaps between the cathode materials, like the gaps between sheets of paper in a book. As mentioned earlier, in cathode materials, metal oxide as acid and lithium ion as alkali neutralize each other and are stable. If a large amount of lithium ions is withdrawn during charging, the neutralized state is disrupted. When the neutralized state collapses, the negatively charged metal oxide layers repel each other, and the structure, which was arranged in an orderly manner like a book, collapses. In addition, the neutralization state with lithium ions is broken and the metal oxide regains its original strong oxidizing property, oxidizing (or burning) the surrounding electrolyte, which can lead to a major fire. These combined factors limit the utilization rate of the cathode.

The utilization rate of the anode side is about 90%, which is higher than that of the cathode. This is because the anode is protected by a film formed by additives. However, if the utilization rate exceeds 100%, the excess lithium ions will combine with electrons and return to metal, and, in the worst case, may break through the separator, leading to an accident. When designing a lithium-ion battery, the voltage should be between 4.2 V and 4.4 V (cathode utilization ratio of 75% and anode utilization ratio of 90%), depending on the material (Figure 3.15).

3.13   Synthesis Methods for Cathode Materials

The first positive electrode material for lithium-ion batteries was theoretically calculated to have a potential of over 4 V, and Dr J. B. Goodenough of the United States and his team synthesized lithium cobaltate.[30] This material is a metal oxide, or ceramic, similar to pottery or porcelain. Since the cathode is basically an oxide ceramic, old knowledge relating to the sintering of ceramic powders has been applied in various ways. Typical cathode synthesis methods include solid-phase, co-precipitation, and hydrothermal synthesis, which can be successfully combined. The solid-phase method is the classical, inexpensive method, in which finely milled solid materials are well mixed and sintered at high temperature. But it is difficult to control the shape and other processing because the shape of the original material is destroyed. In the co-precipitation method, the target metal,

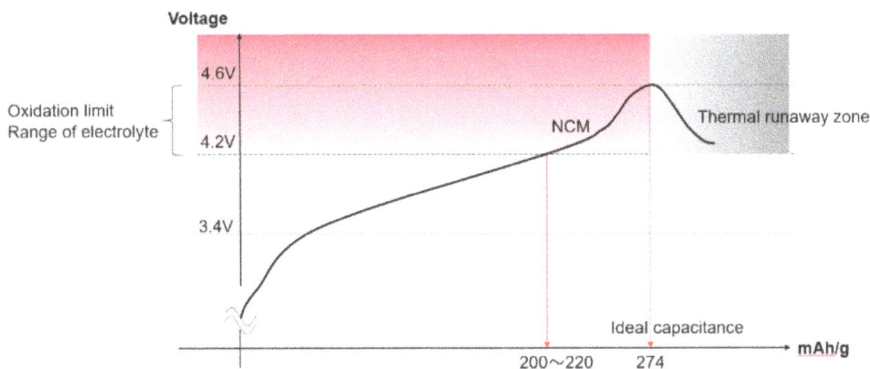

Figure 3.15 Conceptual diagram showing the behavior of the positive electrode of a layered oxide-based lithium-ion battery as it is charged. When electrons move from the positive electrode to the negative electrode, they first detach from the lithium. The lithium then leaves the crystal as lithium ions and moves toward the negative electrode. If the crystal loses too much lithium, it will lose its structure and the crystal structure will change, making it unable to store lithium and reducing its capacity. When the lithium inside the crystal is depleted as the battery is recharged further, the next step is for electrons to be removed from the oxygen inside the crystal. The oxygen, having lost its electrons, leaves the crystal as a radical and, due to its powerful oxidizing power, oxidizes (burns) combustible materials in the surrounding area. This is called thermal runaway.

a strong acid, such as sulfuric acid, and a metal salt are dissolved in water and dropped into an alkaline solution, such as sodium hydroxide, to form a metal hydroxide. This method forms microstructures that are impossible to obtain using the solid-phase method. After forming the precursor using co-precipitation, it must be sintered, as in the solid-phase method.

The hydrothermal synthesis method uses the properties of high-temperature, high-pressure water to dissolve substances that normally cannot be dissolved. The purity achieved with this method is high, and particles of various shapes can be obtained by adjusting the temperature and pressure. For research and development of cathodes, I think it is essential to prepare and combine the above methods.

The cathode materials that can be synthesized by the above method are minute elementary particles and are called primary particles. Particles that can be easily handled as a powder by combining primary particles are called secondary particles. The primary particles must have a shape and fineness that can efficiently adsorb lithium ions, and the surface of the secondary particles must be designed to suppress reactivity by reducing the surface area to suppress electrolyte decomposition.

The tapped density of particles is an indicator of the packing density of secondary particles. The higher the tapped density, the better the packing, which thus increases the volume density of the electrode and the energy density of the battery.

As will be discussed later, once electrodes are made from cathode or anode materials, they need to be painted. It is easier to make a good electrode if it is made into spherical or other secondary particles that are easy to paint in advance.

Secondary particles can be formed by using a special pot, a planetary ball mill, or spray drying.

The hydrothermal synthesis method can design secondary particles in a single step. Furthermore, by strictly controlling temperature and pH during the co-precipitation process, spherical secondary particles can be obtained.

With time and effort, it is possible to create high-performance secondary particles, but the cost is high. It is important to consider matching the cost with market needs.

3.14 Synthesis Methods for Carbon Anode Materials

Oxide-based cathode materials are ceramics, like pottery, so a firing temperature of about 1200 °C is sufficient. This temperature can be achieved with equipment that is easily available online for the hobbyist potter. Synthesizing the anode, especially graphite anodes, requires a higher temperature, and equipment capable of generating temperatures of at least 2000 °C.

Carbon takes many forms, so it is difficult to discuss all the different types of carbon used in the negative electrode of lithium-ion batteries. For this reason, I will focus on graphite-based materials, which are currently the mainstay materials used in lithium-ion batteries. Graphite is a form of carbon. Graphite has a layered structure made up of thin films (graphene) stacked on top of each other. There are slight differences in the way the layers are stacked, and various studies have been conducted to classify these.[31,32]

Artificial graphite is made synthetically, while natural graphite is generated under high temperature and high pressure in the Earth's interior. The crystallinity of graphite tends to increase with the heat (and pressure) applied. The temperatures and pressures in the Earth's interior are tremendous, and graphite with higher crystallinity than artificial graphite is generally obtained. The structure of graphite has been discussed in detail in Section 3.2.

There are two types of carbon materials: easily graphitizable carbon materials and hard graphitizable carbon materials, called soft carbon and hard carbon, respectively.[33,34] Although this is a complicated topic, it should be noted that it does not describe the hardness of carbon. When Dr Yoshino and his colleagues invented the lithium-ion battery, a material based on needle coke,[18] which has lower crystallinity than graphite, was used for the negative electrode. Needle coke is a source of easily graphitizable carbon (soft carbon),[35] which can be easily converted into graphite, and when calcined it becomes artificial graphite. The raw material of needle coke is pitch, a residue obtained from the dry distillation of coal, which contains a large amount of carbon.

Even if the same pitch is used as a raw material, carbon that is passed through pitch-based resins (phenolic resins, *etc.*) is called hard carbon, which does not easily turn into graphite even when synthesized at high temperatures. Carbon that is difficult to graphitize is suitable for synthesizing carbon that contains many voids, and is used as a material for activated carbon, for example, which is used as an electrode material for electric double-layer capacitors. As a side note, carbon with many voids is said to be effective for sodium-ion batteries, where the sodium has a larger ion radius than that of lithium ions. Carbon materials based on hard graphitized carbon sources may be promising in the future.

Let us return to graphite. Typical examples of graphite shapes have been discussed earlier, so I will not discuss them here. As in the case of the cathode, it is necessary to reduce the surface area to suppress secondary reactions while maintaining reactivity to lithium ions, and efforts are being made to make it spherical in shape or to coat the surface with non-crystalline carbon.

When easily graphitizable carbon (soft carbon) is calcined, its crystallinity increases from low-crystallinity carbon to graphite above 2500 °C. Comparing the charge–discharge geometry of graphite and amorphous carbon, graphite has less voltage fluctuation and can have a deeper potential, but amorphous carbon has a lower average battery voltage, which is not favorable from an energy point of view. In addition, a constant battery voltage contributes to overall equipment miniaturization, so the negative electrode of lithium-ion batteries shifted from amorphous carbon to graphite as the IT revolution progressed.

However, even carbon with low crystallinity has sites that can adsorb lithium ions, and there are areas where there are more lithium ion adsorption sites than graphite, and the adsorption potential is slightly higher than for graphite (Figure 3.16), so there is less reduction stress on the electrolyte. Therefore, it has the advantages of superior durability and low-temperature properties. For this reason, blends of amorphous carbon and graphite, or blends of artificial graphite and natural graphite, are used in electrodes for BEVs to improve battery regeneration performance and low-temperature properties.

It has been mentioned that as the firing temperature of artificial graphite is increased, it passes through the amorphous carbon region and becomes highly crystalline graphite (Figure 3.17). I also mentioned

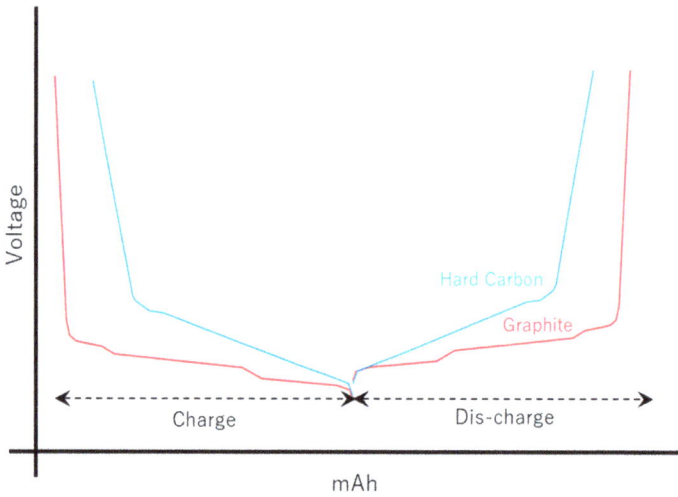

Figure 3.16 Diagram of charge–discharge profiles of hard carbon and graphite electrodes.

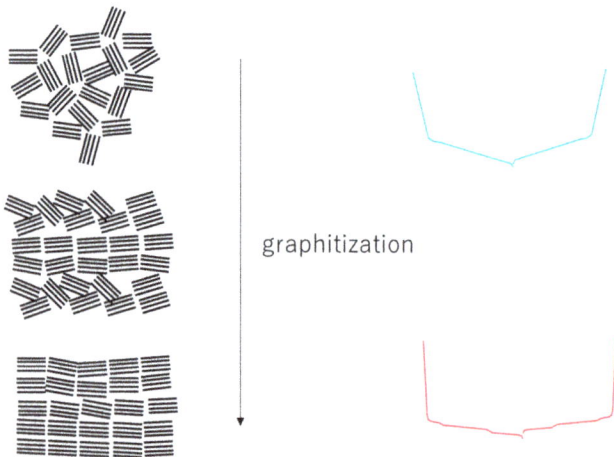

Figure 3.17 Modified Franklin models of graphite/hard carbon.

that amorphous carbon has lower capacitance, but has advantages in low-temperature characteristics and charging performance (regenerative acceptance performance). There was an attempt to produce carbon with a good balance between capacity and durability by adjusting the firing temperature. Mesocarbon microbeads (MCMBs), which were used in the early days, exhibited the same characteristics and were also widely used because they were spherical and very robust. It has recently been attracting attention again as a material for all-solid-state batteries.

References

1. M. Zou, M. Yoshio, S. Gopukumar and J.-i. Yamaki, *Electrochem. Solid-State Lett.*, 2004, **7**, A176.
2. A. R. Armstrong and P. G. Bruce, *Nature*, 1996, **381**, 499.
3. M. Yoshio, H. Noguchi, K. Yamato, J. Itoh, M. Okada and T. Mouri, *J. Power Sources*, 1998, **74**, 46.
4. K. Tatsumi, *Electrochem. Ind. Phys. Chem.*, 1995, **63**, 980, DOI: 10.5796/kogyobutsurikagaku.63.980.
5. A. Satoh, N. Takami and T. Ohsaki, *Solid State Ionics*, 1995, **80**, 291.
6. H. Ago, K. Nagata, K. Yoshizawa, L. Tanaka and T. Yamabe, *Bull. Chem. Soc. Jpn.*, 1997, **70**, 1717.
7. T. Matsumoto, U. Nagashima and K. Tanabe, *J. Comput. Chem., Jpn.*, 2003, **2**, 63.
8. J. O. Besenhard, M. Winter, J. Yang and W. Biberacher, *J. Power Sources*, 1995, **54**, 228.
9. M. Yoshio, H. Wang, K. Fukuda, Y. Hara and Y. Adachi, *J. Electrochem. Soc.*, 2000, **147**, 1245.
10. B. Fuchsbichler, C. Stangl, H. Kren, F. Uhlig and S. Koller, *J. Power Sources*, 2011, **196**, 2889.
11. H. Morimoto, M. Tatsumisago and T. Minami, *Electrochem. Solid-State Lett.*, 2001, **A16**, 72.
12. H. Huang, E. M. Kelder, L. Chen and J. Schoonman, *J. Power Sources*, 1999, **81**, 362–373; E. Feyzi, M. R. A. Kumar, X. Li, S. Deng, J. Nanda and K. Zaghib, *Next Energy*, 2024, **5**, 100176.
13. A. Yoshino, K. Nakanishi and A. Ono, *Jp. Pat.*, 2642206, 1989.
14. M. Ue, *Prog. Batteries Battery Mater.*, 1997, **16**, 332.
15. M. Ue, in *Extended Abstracts of the Battery and Fuel Cell Materials Symposium*, Graz, Austria, April 18–22, 2004, p. 53; Meeting Abstracts of the 12th International Meeting on Lithium Battery, Nara, Japan, June 27–July 2, No. 31, 2004; Meeting Abstracts of the 206th Electrochemical Society Meeting, Honolulu, HI, October 3–8, No. 308, 2004.
16. D. Aurbach, J.-I. Yamaki, M. Saolmon, H.-P. Lin, E. J. Plichta, M. Hendrickson, W. A. van Schalkwijk and B. Scrosati, *Advances in Lithium-Ion Batteries*, Kluwer Academic/Plenum Publishers, New York, NY, 2002, ch. 1, 5, and 11.
17. A. Yoshino, *et al.*, *Jp. Pat.*, 2128922, 1984.
18. A. Yoshino, *Lithium-ion Battery Story – In The World of Japanese Technology Break*, CMC, Japanese edn, 2004, pp. 58–62.
19. H. Yoshitake, K. Abe, T. Kitakura, J. B. Gong, Y. S. Lee, H. Nakamura and M. Yoshio, *Chem. Lett.*, 2003, **32**, 134.
20. P. Ghimire, H. Nakamura, M. Yoshio, H. Yoshitake and K. Abe, *ITE Lett. Batteries, New Technol. Med.*, 2005, **6**, 16.

21. P. Ghimire, H. Nakamura, M. Yoshio, H. Yoshitake and K. Abe, *Electrochemistry*, 2003, **71**, 1084.
22. M. Fujimoto, *et al.*, *Jp. Pat.*, 3059832, 1992.
23. N. Dimov, S. Kugino and M. Yoshio, *Electrochim. Acta*, 2003, **48**(11), 1579.
24. V. Ponnuchamy, S. Mossa and I. Skarmoutsos, *J. Phys. Chem. C*, 2018, **122**(45), 25930–25939.
25. K. Dokko, D. Watanabe, Y. Ugata, M. L. Thomas, S. Tsuzuki, W. Shinoda, K. Hashimoto, K. Ueno, Y. Umebayashi and M. Watanabe, *J. Phys. Chem. B*, 2018, **122**(47), 10736–10745.
26. J. B. Goodenough and K. Mizuchima, *US Pat.*, 4302518, 1980.
27. X. Yang, X. Sun, M. Balasubramanian, J. McBreen, Y. Xia, T. Sakai and M. Yoshio, *Electrochem. Solid-State Lett.*, 2001, **4**, A117.
28. Y. Xia and M. Yoshio, *J. Electrochem. Soc.*, 1997, **144**, 4186.
29. *NEDO Rechargeable Battery Technology Development Roadmap*, Japanese edn, 2013, https://www.nedo.go.jp/
30. K. Mizushima, P. C. Jones, P. J. Wiseman and J. B. Goodenough, Li_xCoO_2 (0<x<-1): A new cathode material for batteries of high energy density, *Mater. Res. Bull.*, 1980, **15**(6), 783–789.
31. R. E. Franklin, *Acta Crystallogr.*, 1951, **4**, 235.
32. J. C. Bowman, *Proc. 1st and 2nd Conf. on Carbon*, Buffalo University, 1956, vol. 59.
33. S. Ergun, *Carbon*, 1968, **6**(2), 141.
34. R. E. Franklin, *Acta Crystallogr.*, 1950, **3**, 107.
35. A. Yoshino, *Lithium-ion battery story - in the world of Japanese technology break*, CMC, Japanese edn, 2004, pp. 54–62.
36. https://www.ites.co.jp/section/index/processing_technologies/li-ion-battery_analysis.html.

4 Li-ion Battery Manufacture: Preparation of Electrodes

4.1 Introduction

In this chapter, I explain how lithium-ion batteries are produced based on my own experience. The manufacturing process for cells is divided into two main stages: the process of turning materials into electrodes and the process of assembling the electrodes with other components. In this chapter, I explain the process of making electrodes. The electrodes are the key to the battery, and if they are not made properly, all subsequent efforts will be in vain. The assembly process after the electrodes is explained in Chapter 5.

When referring to lithium-ion batteries, small individual batteries are also referred to as batteries, as are the batteries installed in battery electric vehicles (BEVs). In the battery industry, the smallest unit of a battery is called a battery cell and the largest mass is called a battery pack, to avoid confusion. A module is a mass of cells that are combined, to a certain extent. So, in summary, from smallest to largest we have: cell → module → pack (Figure 4.1). Modules and packs are designed by final product manufacturers for customers for their own products, so I will not go into detail about them. However, I would like to discuss some key design points.

RSC Foundations No. 4
Li-ion Batteries
By Hitoshi Nakamura
© Hitoshi Nakamura 2025
Published by the Royal Society of Chemistry, www.rsc.org

Figure 4.1 Internal photograph of a commercially available battery pack. This is a first-generation pack structure for electric vehicles, and the stages of cell → module → pack are relatively clearly visible. As there are extra parts that mainly correspond to module components, it is thought that there will be a gradual shift to Cell to Pack, where the cells are directly arranged in the pack, and to Cell to Chassis, where the cells are directly arranged in the chassis, as seen from the vehicle, as there are extra parts that construct the pack structure. (The author took these photos during a disassembly survey for educational purposes.)

As an example, I will briefly touch on the packs for automobiles. BEVs are limited in terms of the volume and surface area they can carry, so there is always a demand to reduce the number of peripheral components used to make up cells, modules, and packs as much as possible. In the past, batteries for mobile devices were used as is, or in larger sizes, so many support components were used to make the battery cell into a pack. In a battery, the active materials of the positive and negative electrodes are the only parts that generate capacity, so the volume and weight of the collecting electrodes, separator, and packaging materials, for example, should be minimized. The larger the cell, the more advantageous it is in terms of volume (weight) and energy density, since the volume of the packaging material can be minimized by making the cell larger. Recently, the concepts of Cell to Pack (CTP) or Cell to Chassis (CTC) have been actively developed, in which a large cell designed with the vehicle layout in mind is placed directly on the vehicle to minimize unnecessary components and maximize the electric energy that can be installed. With regard to batteries for BEVs, the trend is towards integrated development of the battery and vehicle. For this reason, automakers that cannot manufacture their own batteries or participate in the design of their own cells will be at a disadvantage in terms of cost and performance.

4.2 Preliminary Process – Preparation of Electrodes

4.2.1 Electrodes Are the Most Important Element of a Battery

The quality of the electrodes has a decisive influence on the performance of the battery. If the electrodes are not precisely made, it is useless, no matter how hard you work in the subsequent processes.

4.2.2 Why Thin-film Electrodes for Lithium-ion Batteries?

Water has been used as the electrolyte in classical batteries, such as lead-acid batteries. Compared to water, the organic solvents used in lithium-ion batteries have high viscosity, making it difficult for lithium ions to move and reducing their migration speed. In lead–acid batteries, hydrogen moves. The difference in migration speed between hydrogen in lead–acid batteries and lithium ions inside lithium-ion batteries is approximately 100 times greater.[1,2] Therefore, in lithium-ion batteries, it is necessary to make the electrodes thin to shorten the distance ions must travel. This allows a power output equivalent to that of a lead-acid battery to be obtained. However, the ratio of components other than the cathode and anode materials increases, so although the power increases, the amount of energy stored decreases. Electrode thickness is designed according to the application of the battery. Lithium ions emitted from the electrode move through the electrolyte and the electrode must contain a suitable amount of electrolyte that acts "like a sponge". As the electrode density increases, the amount of electrolyte that can soak into the electrode decreases, which reduces the path for lithium ions to flow through, resulting in a loss of power and a decrease in battery capacity. Therefore, it is important to keep in mind that the density of the electrode has a specific impact on the power. My research has not revealed who first developed the technology to make thin films for lithium-ion battery electrodes. However, the Yoshino group, which later won the Nobel Prize, faced significant difficulties in producing thin films for electrodes.[3] It is said that many

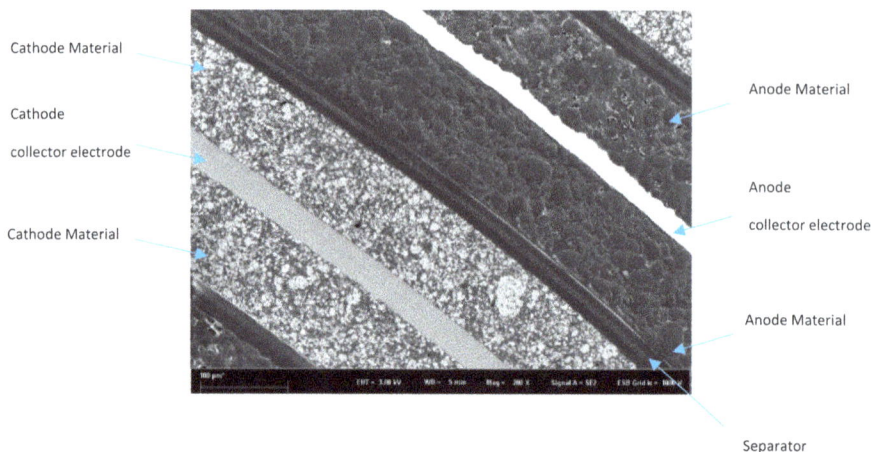

Cathode Material

Cathode

collector electrode

Cathode Material

Anode Material

Anode

collector electrode

Anode Material

Separator

Figure 4.2   An example of a scanning electron microscopy (SEM) cross-section image of the configuration of the cathode, anode, and separator of a lithium-ion battery.[6] Reproduced from ref. 6 with permission from ITES Corporation.

Japanese companies were able to enter the lithium-ion battery industry because they possessed mass production technologies used in magnetic recording tapes. As will be discussed later, it is natural to consider the technology for dispersing various slurries as an application of magnetic tape technology. Figure 4.2 shows a cross-sectional photograph of the basic elements of a battery, which consists of a thin positive and negative electrode separated by an electronically insulating separator and a current-carrying current collector.

4.2.3   Basic Structure and Manufacturing Method of Electrodes for Lithium-ion Batteries

The cathode electrode for lithium-ion batteries consists of a thin film of aluminium (about 5 to 20 μm thick) coated onto both sides with a 50 to 200 μm thick film of a material (called a cathode compound) made of a mixture of cathode material, conductive carbon, and a small amount of adhesive. The anode consists of a thin copper film (5–20 μm thick) coated on both sides with a thin film of graphite mixed with adhesive (called an anode compound) at a thickness of 30–120 μm. The weight ratio of the active material layers of the positive and negative electrodes needs to be determined based on the ratio of the capacity of the negative electrode to the capacity of the positive electrode. The density of the positive and negative electrodes affects the lifespan and output of the battery, so it is necessary to set the density appropriately. Generally, the standard ratio of the capacity of the positive and negative electrodes is around $1:1.1$. However, the optimal value differs depending on the material used, so it is necessary to check this through experiment, such as by making small cells. The density of the active material layer in the positive electrode is around $2-3$ g cm^{-3}, and that of the negative electrode is around 1.5 g cm^{-3}. If the ratio of the positive electrode to the negative electrode is not correct, this will have a serious impact on safety. For example if the anode is thinner than designed, the absorption limit of lithium ions in the anode is exceeded and dendrites of metallic lithium occur, which in the worst case may break through the separator and induce a short circuit, leading to an accident. Dendrite occurrence has a critical impact on battery life. If the anode is relatively too thick, excessive solid–electrolyte interface (SEI) formation on the carbon surface will eat away lithium ions that can move in the electrolyte, again affecting the life of the battery. The graphite anode has a higher capacity for its mass than the cathode, so it must be finished more precisely. In my experience, the allowable coating weight error of the electrode is $\pm 0.5\%$, or less, of the design value ($\pm 1\%$ or less for the cathode and anode combined).

4.2.4   Selection of the Type of Coater

The quality of the electrodes has a significant impact on their capacity, output, lifespan, and safety of the battery, so it is important to carefully choose the coater that is applied to the electrodes. There are various types of coaters used to coat the collecting electrode. The coater used for lithium-ion batteries is mainly a roll coater. The collector electrode is moved and the slurry is applied to the surface. There are also different coating head types, for example, comma, comma reverse, and die head. The die head is a coater that continuously realizes precise coating, as described in Section 4.2.3. A simple mechanism is shown in Figures 4.3 and 4.4 and Video 4.1, https://www.youtube.com/watch?v=r4ON5KjTo8s.

4.2.5   Overview of Electrode Manufacturing Process

The cathode or anode active material, conductive material, and adhesive are mixed to form a paint-like coating (called a slurry or coating liquid), which is then applied thinly and evenly using the coating equipment. After the coating liquid is degassed to prevent bubbles from forming, it is pressurized and sent to the die head of the coating

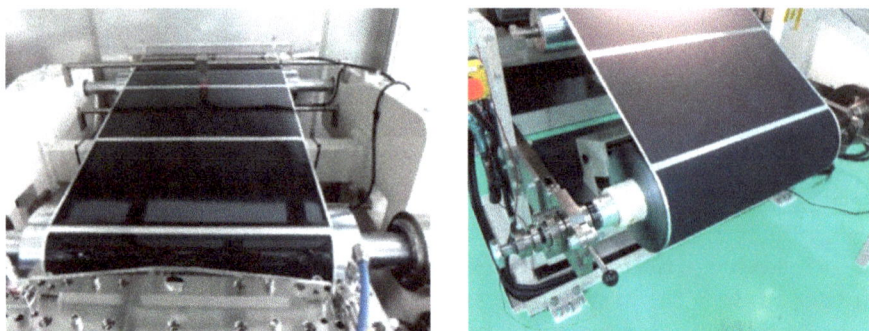

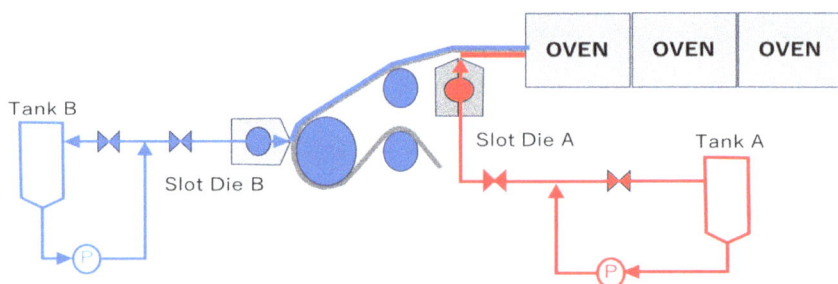

Figure 4.3 Photo of a coating machine (cathode side) and diagram of a typical electrode coating system with a slot die.

Figure 4.4 Photograph showing the coating of the negative electrode by a roll coater.

machine by a pulsation-free pump called a mono-pump. During the process, the liquid passes through a filter to remove magnetic metals and fine lumps. The slurry is a thixotropic fluid, which means that when it is pushed out of the die head, pressure is applied to reduce viscosity and ensure uniform coating. Drying conditions are also very

Figure 4.5 A drying furnace of a coating machine that I often used for prototyping. This furnace is capable of drying substrates up to 750 mm wide. The furnace is 25 m long and divided into eight zones. It is set up as a large pilot machine rather than as a mass production machine.

important. Direct drying of the slurry or drying at too fast a rate can cause concave, convex, or high edges, in which the edges of the coating film are raised. In the initial stage of drying, it is necessary to dry the slurry at a low temperature and in a mild manner. Die-head coaters have a multi-stage drying chamber. It is usually divided into 4–6 zones (Figure 4.5). Know-how is required to gradually increase the air volume and temperature. When the solvent is almost dried, the output is turned up to the maximum and the drying process is completed. If the slurry does not dry, problems will occur, and, in the worst case, the foil will have to be replaced over a distance of more than 200 meters. I have experienced this situation several times, and it is quite time-consuming and disheartening. This is a part of the process that can only be understood through actual coating experience and requires craftsmanship. Once dried, the material is turned over and coated on the reverse side in the same way. Recently, it has become possible to coat both sides at the same time, but when I was involved in the manufacturing process, one-sided coating was the norm due to coating accuracy issues. The electrode drying process in the manufacture of lithium-ion batteries consumes a large amount of energy, so simultaneous coating is advantageous if possible. In any case, I recommend single-sided coating for the initial condition setting, as everything is done step by step.

After coating, the electrodes are pressed in a large press called a roll press to achieve the desired density. Generally, electrodes become softer when heated, so they are preheated before pressing. The density of the anode is very important, and if the density is too high, it may affect the movement of lithium ions in the electrolyte, resulting in an extreme drop in power output. Especially at low temperatures, the output may be extremely reduced, so it is necessary to adjust the density according to the operating environment. When aiming for a density of approximately 1.5 g cc^{-1} or higher, it is advisable to investigate the temperature dependence of the prototype battery. The density of the cathode also affects output, but not as significantly as that of the anode. The density to aim for is around 2.8–3.2 g cc^{-1} for the cathode. Since the electrode density affects the volume energy density of the battery, it is better to have a high electrode density, but the above considerations must be kept in mind. In any case, it is necessary to confirm the output characteristics of the battery by making a prototype.

Gap rollers are not suitable for precision pressing, and rollers that use hydrostatic pressure should be used. The precision of the electrode also depends on the roller diameter, so to achieve a higher density, a larger roller diameter tends to be needed. In general, roller systems that enable high-density electrode formation with small-diameter rollers are superior and should be used as a guideline for selection.

Next, the pressed rolls are dried in a dryer. Vacuuming is very effective at this point, but vacuuming does not transfer heat, and the drying process takes a long time. It is more efficient to vacuum the rolls and then heat them by introducing inert gas.

The die-head coater discussed above was originally used in the textile industry for mass production of resin films and this die-head coater is capable of coating in a variety of patterns, with typical examples being intermittent coating and stripe coating.

Intermittent coating is used when the uncoated area is horizontal to the direction of the electrode. Stripe coating is a method of manufacturing electrodes in which uncoated areas are created parallel to the direction of electrode coating. The introduction of intermittent coating requires a reduction in the electrode manufacturing speed, and the stripes tend to wrinkle during pressing, so it is necessary to design cells with uncoated areas in mind.

Next, there is the slitting process. When I was involved in mass production, the roller cutter method, which uses two blades to cut the electrode, was the main method used. In the slitting process, burrs generated from the electrode cutting surface are the biggest concern.

Figure 4.6 Cross-section of a slit anode. If metal burrs appear on the cross-section of the electrode, they will break through the separator when the battery is assembled and cause a short circuit. In the case of a slitter using a metal blade, burrs are more likely to appear if the blade becomes dull. The presence of burrs should be controlled by determining the allowable slot time in advance, or by pulling out the foil for each slot and observing the cross-section.

For this reason, detailed blade control was extremely important, such as pulling out the electrode, observing the cross-section, and controlling the number of shots of the blade (Figure 4.6). Recently, a method of cutting using a laser has been developed, and the complication of burrs seems to have been eliminated. In some cases, the slitting process is carried out before the drying process to save drying time. This is a matter of choice for the factory designer. Of course, the fewer the number of processes, the lower the cost.

4.2.6 About the Coating Slurry

From my time involved with the mass production of cells, I had the following understanding: assuming that there were no problems with materials, 70% of battery performance was determined by the electrode and 70% of the electrode performance was determined by the coating solution. Therefore, the adjustment of the coating solution requires a great deal of know-how. The coating solution consists of cathode or anode material, adhesive (called binder), conductive auxiliary agent (acetylene black, fine carbon, crushed graphite, *etc.*), and solvent. The following is a typical recipe for cathode production. The ratio of cathode, conductivity aid, and binder is roughly $95:3:2$. The solid concentration should be 65% to 75% and the viscosity should be within the range of 5000 to 10 000 mPa s (5000 to 10 000 mPa s is about the viscosity of honey or water syrup) (Video 4.2, https://www.youtube.com/watch?v=BKMoplIMtQw).

The thicker the solid concentration, the better, as long as it is below the target viscosity. This is because when the electrode is dried in the drying process, evaporation of the solvent may affect the smoothness of the electrode. In other words, it is part of the developer's skill to make a slurry with high concentration and low viscosity. First, the cathode and conductive auxiliary agent are mixed in a powder mixture. Next, appropriate amounts of adhesive and solvent are mixed and stirred to form a viscous compound. The viscous compound is then kneaded well to uniformly disperse the powders. Kneading the viscous compound to obtain a high degree of dispersion is called "hard kneading". The kneaded compound is dissolved by pouring a solvent into it. This is sufficient, but if the viscosity is high, a device called a disperser is used (Figures 4.7 and 4.8).

Historically, there have been two main methods of producing slurry for lithium-ion batteries. One is to create a uniform slurry with few lumps by dropping a mixture of active material and conductive auxiliary agent into a viscous solution, and the Eirich mixer is suitable for this (Figure 4.9). It is an excellent mixer capable of dispersing materials without subjecting them to stress.

The method of hard kneading using a planetary mixer (Figure 4.7) has been used for a long time. When used by skilled craftsmen, hard kneading can produce a slurry that is evenly dispersed. Since intense shear stress with high viscosity is applied, it is easy to obtain a uniform

Figure 4.7   Planetary mixer. It has been known since ancient times that kneading with very high viscosity is effective in forcing particles to disperse. For example, when you knead bread, do you knead the dough in a loose state with a lot of water? I am sure that you knead it like clay, without much water. This is called hard kneading. To achieve this, mixers with powerful motors and sturdy blades were developed. The blades have a planetary motion. Some are equipped with high-speed blenders for further dispersion after kneading. Photo courtesy of Primix Japan.

and highly concentrated slurry with few lumps. However, if care is not taken, stress is placed on the material. In recent years, there have been many good dispersing machines available, so it may be convenient to use this mixer for rough kneading and then use a dispersing machine to disperse the material at once. Primix's Fillmix (Figure 4.10) or Bead mill (Figure 4.11) are typical examples of such dispersers.

Adhesives (binders) are used for both positive and negative electrodes. For the positive electrode, polyvinylidene fluoride (PVDF) is used, which has been used since the first lithium-ion battery due to its strong resistance to oxidation and high affinity for electrolyte solution. N-Methyl-2-pyrrolidone (NMP) is used as a dilution solvent for the coating solution. Styrene butadiene rubber or modified nitrile rubber emulsions are used as anode coating solutions. The viscosity of emulsions, which are aqueous solutions (or, more correctly, dispersions),

Figure 4.8 Finished view of slurry mixed with a planetary disperser. The slurry is very fine using so-called solid kneading. The container is 100 L, and is the type used for prototype production. Planetary mixers (with disperser) that can perform solid kneading require extremely high motor output. Since solid kneading inevitably places stress on raw materials, the most common method these days is to use this type of mixer for rough kneading and then run it through a disperser for final adjustment.

Figure 4.9 Eirich mixers have been used since the early days of lithium-ion battery production. The basic idea is to feed material into a viscous fluid little by little to create a uniform slurry without stress to the material. This mixer is also used for granulating active materials, making it an all-round mixer. Photo courtesy of Eirich Japan.

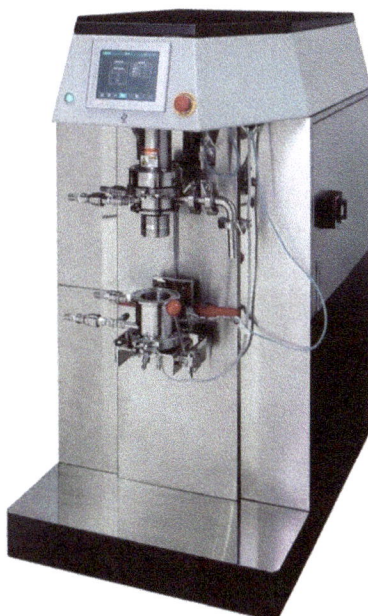

Figure 4.10 Fillmix is a unique dispersion device. Shearing the slurry into thin layers has the effect of reducing viscosity without changing the concentration of the slurry. Also, since the machine does not directly touch the area where shearing is applied, there is little adverse effect on the material. I have used this method myself. Photo courtesy of Primix Japan.

is lower than that of PVDF/NMP slurries and may not easily reach the viscosities required. To compensate for this, carboxymethylcellulose (CMC) is used as a thickening agent. First, the active material is kneaded with an aqueous solution containing CMC. After kneading, water is added to adjust the viscosity to the optimum level. The pH of the water is very important, so all water used must be pure water or ion-exchanged water. Finally, mixing the binder, styrene butadiene rubber/nitrile butadiene rubber, may cause the viscosity to be lower than intended, so it is necessary to repeat preliminary experiments to optimize the recipe so that the slurry viscosity is optimal after mixing the binder dispersion solution.

For both positive and negative electrodes, the viscosity of the slurry will fluctuate over the course of half a day to a day. A well-made slurry will shift to a lower viscosity. The coating process should be conducted with this viscosity movement in mind. The period during which a slurry retains its properties without settling is called the pot life. The pot life of the slurry should be at least one week. It is ideal to have

Figure 4.11 A bead mill is a wet media dispersing machine that pulverizes and disperses particles in a slurry by agitating the media, called beads, which are packed in the grinding chamber, using the action of shear and impact between the beads. Bead mills are capable of producing fine particles down to the sub-micrometer to nanometer level, but they are also difficult to set up because of the stress they put on the material. They are used when you want to produce fine particles down to the sub-micrometer level, such as conductive additives, or when you want to mix active materials at the sub-micrometer level. Photo courtesy of AIMEX CO., Ltd.

separate facilities for producing aqueous and NMP-based slurries, as contamination whereby the positive electrode material mixes with the negative electrode material must be avoided, and the NMP/PVDF solution becomes gelatinous and viscous when water is mixed with it.

Incidentally, have you ever experienced trying to dissolve CMC but finding it took a long time to dissolve? The JetPaster mixer (Figure 4.12) has been used to dissolve CMC in water and to disperse a lithium-ion battery slurry in water or organic solvent in a short period of time.

4.2.7   Next-generation Electrode Coating Methods

NMP is currently commonly used as a solvent in the slurry for the positive electrode. This is because the positive electrode active material is made from layered oxides, which have been used since the early days

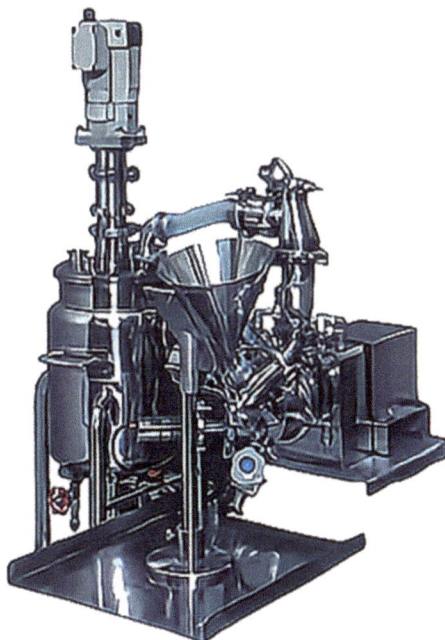

Figure 4.12 JetPaster is a product name. The official name is "Powder absorbing dissolving and dispersion device", and this is a unique mixer. Although the manufacturer of the JetPaster does not disclose details of the mixing principles, it uses the cavitation phenomenon. A high-speed rotor will generate very low pressure to create cavitation bubbles. The cavitation bubbles contribute to dissolving CMC in water and disperse agglomerated lithium-ion battery powders by expansion and shrinkage of the cavitation bubbles. Photo courtesy of Nihon Spindle Manufacturing Co., Ltd.

of lithium-ion battery production, and are vulnerable to moisture. When it comes into contact with water, lithium dissolves from within the active material, and, in the worst-case scenario, the crystal structure collapses. There is also the issue that lithium ions that dissolve in water change the aqueous solution to a strong alkaline state, causing the aluminium current collector to dissolve. However, the lithium iron phosphate cathode material that has been used extensively in recent years has relatively robust crystals, so even if water is used as the solvent, it is possible to obtain a good electrode by controlling the pH appropriately, and so there are an increasing number of manufacturers using aqueous slurries. Depending on the cathode material used, it is possible to use aqueous solvents that are harmless to the environment and the human body.[4,5]

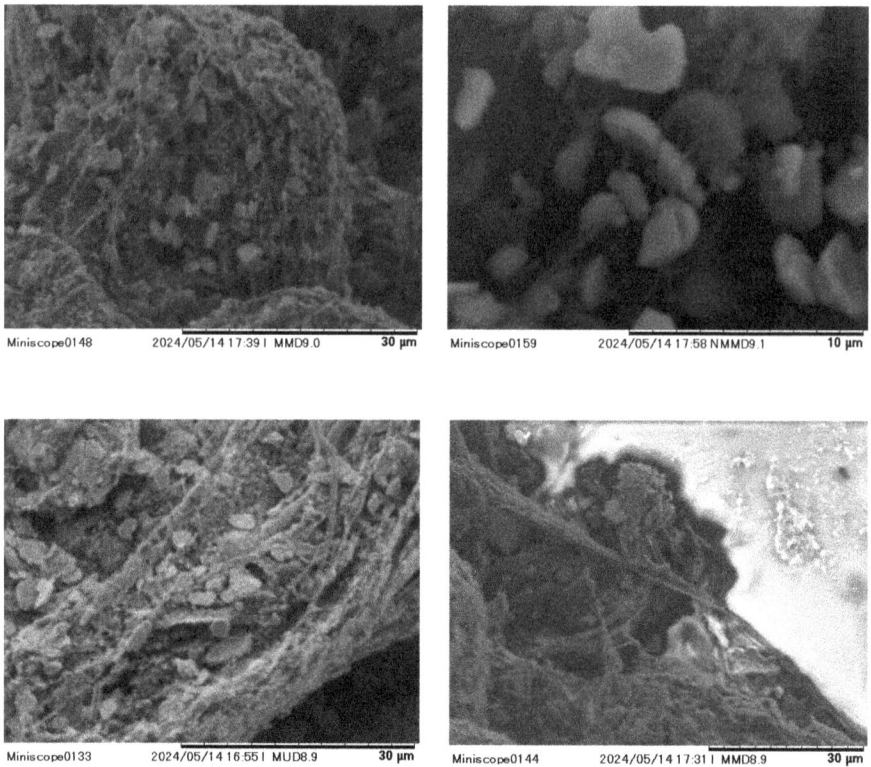

| Miniscope0148 | 2024/05/14 17:39 I MMD9.0 | 30 μm | Miniscope0159 | 2024/05/14 17:58 NMMD9.1 | 10 μm |

| Miniscope0133 | 2024/05/14 16:55 I MUD8.9 | 30 μm | Miniscope0144 | 2024/05/14 17:31 I MMD8.9 | 30 μm |

Figure 4.13   Although coolly called a "dry electrode", the method of electrode formation that takes advantage of the fact that PTFE becomes fibrous when shear stress is applied has been in use for at least 30 years. Electron micrographs show the PTFE transition entangling the granular active material.

The pressed-powder method has been known for a long time as an electrode molding technique, and it uses polytetrafluoroethylene (PTFE or Teflon) as an adhesive. When Teflon and cathode material are mixed, the Teflon becomes fine fibers, and these fibers bind the powder together (Figure 4.13). The powder is then stretched with a roller at the appropriate point, and pressure is applied to attach it to a collecting electrode that has been coated with conductive adhesive in advance. PTFE cannot be used for the negative electrode, since it undergoes reductive decomposition at a potential of 1 V or less. Since no solvent is used, the drying process is simplified, leading to energy savings during battery production, and thick, dense, and uniform electrodes are easily obtained. However, PTFE is not suitable for producing uniform electrodes at high speed because the slightest shear would cause a sudden change in the state of the PTFE. The photographs in

Figure 4.13 show SEM images of the threaded structure of PTFE. The PTFE compacting method has long been used in electric double-layer capacitors as a technology for forming electrodes. Another method is called the poultice method, which can be used for both positive and negative electrodes. The cathode material and adhesive are made into a coating solution using electrolyte. This is applied thinly to the collecting electrode in a dry atmosphere and immediately assembled. The electrolyte is applied like a poultice, which allows the electrode to be assembled in the subsequent steps. There are many secondary benefits, such as the fact that good battery characteristics can be obtained because the electrolyte is spread to every corner of the electrode from the outset.

References

1. E. Logan, E. Tonita, K. Gering, J. Li, X. Ma, L. Beaulieu and J. Dahn, A Study of the Physical Properties of Li-Ion Battery Electrolytes Containing Esters, *J. Electrochem. Soc.*, 2018, **165**, A21–A30.
2. L. Arenas, F. Walsh and C. Ponce de León, The Importance of Cell Geometry and Electrolyte Properties to the Cell Potential of Zn-Ce Hybrid Flow Batteries, *J. Electrochem. Soc.*, 2015, **163**, A5170–A5179.
3. A. Yoshino, *Lithium-ion battery story - in the world of Japanese technology break*, CMC, Japanese edn, 2004, pp. 84–87.
4. M. Kodo and N. Ohnishi, Yamaha Technical review, 2014, https://global.yamaha-motor.com/jp/design_technology/technical/publish/pdf/browse/50gs05.pdf.
5. N. Aguiló-Aguayo, D. Hubmann and F. U. Khan, *et al.*, Water-based slurries for high-energy LiFePO$_4$ batteries using embroidered current collectors, *Sci. Rep.*, 2020, **10**, 5565.
6. https://www.ites.co.jp/section/index/processing_technologies/li-ion-battery_analysis.html.

5 Li-ion Battery Manufacture: Cell Shape and Assembly Method

Following on from Chapter 4, this chapter explains the battery assembly process from electrode formation onwards. The battery assembly process differs slightly depending on the shape, and so is explained separately for each shape. There are three main shapes of lithium-ion battery: cylindrical, prismatic (square), and pouch-type. Each type has its advantages and disadvantages. These are explained in detail in this chapter.

5.1 Assembly Process and Cell Shape

This section briefly describes typical cell shapes and assembly methods. There are three main types of lithium-ion battery cells: cylindrical, square, and pouch (Figures 5.1 and 5.2).

The cell assembly process differs depending on the cell shape. Here, I will discuss the points to note for each of the typical cell shapes shown in Figure 5.2.

RSC Foundations No. 4
Li-ion Batteries
By Hitoshi Nakamura
© Hitoshi Nakamura 2025
Published by the Royal Society of Chemistry, www.rsc.org

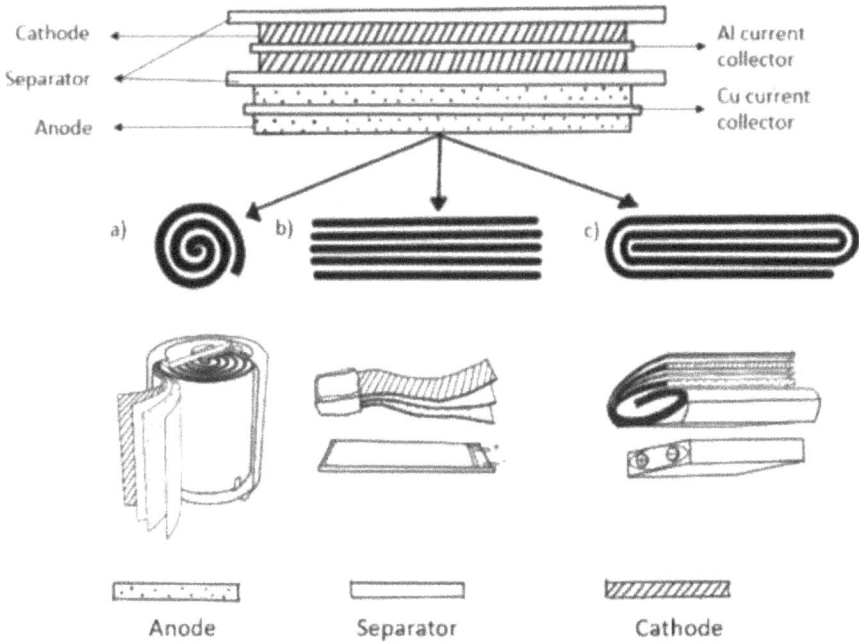

Figure 5.1 A simplified lithium-ion battery configuration. I previously asked LinkedIn readers if they would be willing to draw a simple diagram illustrating a typical cell type. One reader made this nice illustration; it is simple but illustrates all the necessary features and is easy to understand. Illustration courtesy of Dr Siri Gani, Technical University of Darmstadt.

Figure 5.2 Historically, there are three types of cell shape: cylindrical, square, and pouch. The contents of cylindrical cells are jelly-roll shaped, while the contents of square- and pouch-shaped cells are of two types: oval jelly-roll shaped and laminated. Illustration courtesy of Glimpse Engineering, Inc.

5.2   Cylindrical Type

Inside the cylindrical type are alternating rolls of positive and negative electrodes and separators. This is called a wound body (or "group
winding" element). Lithium-ion batteries were the first cylindrical
batteries to appear on the market. Sony developed the first cylindrical
lithium-ion battery in 1991 for the Handycam, which was designed
to be compatible with the nickel metal hydride batteries used in the
Handycam, and developed the so-called 18650 type, a cylindrical battery with a diameter of 18 mm and a length of 65 mm. The 18650
type has long been recognized as the *de facto* standard for lithium-ion
batteries, and with the IT revolution, the 18650 type was widely used
in notebook PCs. Later, the output was increased and used in various other areas, such as power tools. As is well known, a large number of 18650-type cells were used in Tesla's early vehicles. In recent
years, the demand for higher capacitance and increased output has
led to the introduction of the slightly larger 21700 type, which has
been developed for use in mobile devices. However, it is believed to be
relatively difficult to extract the output from cylindrical-type batteries
because of the narrow terminals for extracting current. As objective
evidence, when disassembling used 18650 cells, it is common to see
the electrodes around the extraction terminals deteriorating. One of
the reasons for this is that the resistance around the terminals is high
and generates a lot of heat. Degradation of the battery as a whole is
triggered by the progression of such partial degradation.

Cylindrical batteries have a long history of use in IT portable equipment and are basically energy density oriented. For this reason, a
high electrode density tends to be used, resulting in a relatively small
amount of electrolyte contained within the electrodes. The same tendency is seen in the 21700 model, and it is not known at this time
whether this tendency has been corrected in the 4680 model. It will be
necessary to benchmark the results for this battery in the future. Lithium-ion batteries start to deteriorate if there is an imbalance in the
electrode thickness or heat concentration, *etc.*, which is the starting
point of deterioration. Assuming that the basic design is established,
the three principles for manufacturing and utilization to obtain long-
life batteries with excellent degradation characteristics are as follows:

- use of homogeneous materials
- use of homogeneous electrodes
- homogeneous battery assembly.

The key to obtaining long-life battery packs is the technology to ensure that all batteries are used under homogeneous conditions and that deterioration is uniform. It is important to uniformly distribute the current and manage the heat so that the temperature of the entire battery pack is as uniform as possible. Cylindrical batteries are ideal in shape because they can be wound while maintaining a constant tension applied by the separator. Basically, the winding, or assembly speed, is also fast. Figure 5.3 shows a winding device for the most advanced cylindrical element.

In general, for cylindrical batteries, heat tends to accumulate in the winding core and the effective volume of the drawer terminals is relatively small compared to the area of the electrodes; thus, the current is concentrated in that area and generates heat. This is the starting point for the deterioration of the battery. In addition, as mentioned above, lithium-ion batteries for IT mobile applications, although they have a high energy density, tend to have a high electrode density and relatively little electrolyte, so they tend to deteriorate more easily than batteries made specifically for BEVs. Cylindrical batteries have a limited area to draw current from, making it difficult to extract the output, and as a result they generate a lot of heat and deteriorate quickly, which has been a drawback since their development. An ingenious way to compensate for such weaknesses of cylindrical batteries is to make them larger and to use the concept of "all tabs", in which the

Figure 5.3 State-of-the-art winders for lithium-ion batteries. Winding, insulation, and other processes to prevent short circuits are carried out simultaneously at high speed. Photo courtesy of KAIDO Manufacturing.

current is drawn from the entire electrode. Cylindrical batteries tend to trap heat at their center, but adopting an "all-tab" design improves this issue. Specifically, the positive and negative electrodes are wound vertically, and a large current is drawn from the large area of the current-collecting electrode surfaces created at the top and the bottom. In this way, the area from which current can be drawn is expanded and homogenized, so that overall heat generation can be suppressed, and there is less chance that deterioration due to heat generation in a specific area will spread to the entire battery. This technology has been adopted by Tesla as the basic technology for the 4680 cylindrical type (Figure 5.4).

The all-tab method itself was an early technology established for electric double-layer capacitors, which require instantaneous high current extraction. The earliest cylindrical capacitors with the same structure were used in Honda fuel cell vehicles (FCVs).

The method of forming elements by winding was originally established for aluminium electrolytic capacitors, and the long history of the development of winding machines means that it is used to this day as it is more advantageous in terms of manufacturing cost than other shapes.

The outer case for the batteries is made of iron formed into a cylinder and plated with nickel. The cylinders were developed by a small company in downtown Tokyo, using a traditional Japanese technique

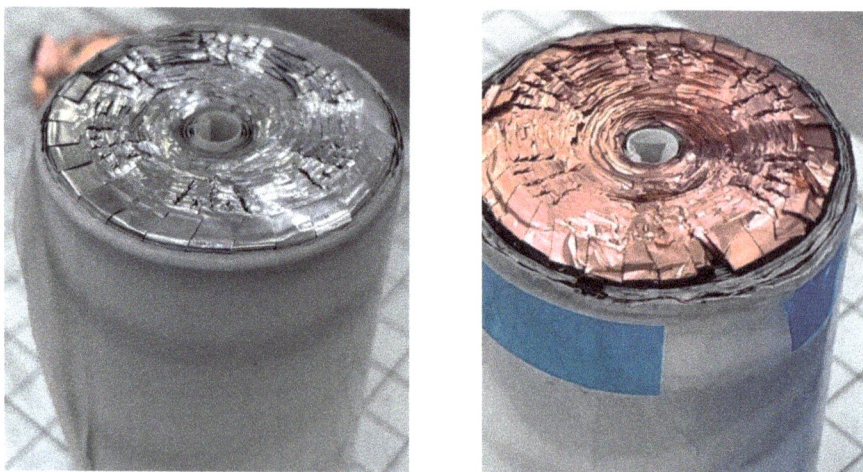

Figure 5.4　Photograph of an exploded view of a commercial cell that uses a double (all)-tab structure. The structures of the collecting electrodes protruding from the positive and negative electrode sides are neatly folded.

called deep drawing. The process of gradually forming a disk into a cylinder is a sight to behold, but I will not go into detail here. After placing the coil-like body called a "jelly roll" in the cylinder and welding the negative terminal to the bottom of the can, a concave surface is made at the top of the cylinder (this process is called necking or beading) and a plastic part of the cylinder made of polypropylene, called the "top insulator", is placed over the concavity. Since the case is basically a negative terminal in the cylindrical type, the top insulator serves to insulate the positive and negative terminals, and, at the same time, it seals the space between the cylindrical case and the cap to prevent leakage of the contents. Next, the positive terminal is welded to the cap. Once this is done, electrolyte is injected through the gap between the cap and the cylinder, and the electrolyte is soaked well by repeatedly drawing a vacuum and applying pressure (this process is called "soaking"). Once the electrolyte is impregnated, the cap is inserted, and the top of the cylinder is woven inward to seal the cell (Figure 5.5).

The completed cell is then initialized by applying voltage, aged, and finally checked for capacitance, resistance, and open circuit voltage (OCV). Initialization and aging will be explained later in this chapter (Section 5.5). In the case of cylinders, there is no need to vent the gas generated by the initialization process. The cylinder is a kind of pressure vessel, and the initial outgassing is contained inside. The cap has

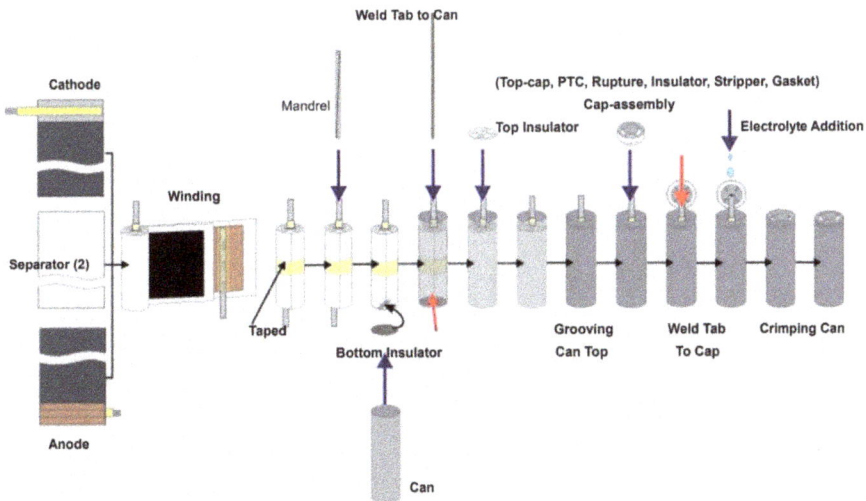

Figure 5.5 Assembly flow diagram for a cylindrical lithium-ion battery. Illustration courtesy of Hohsen.

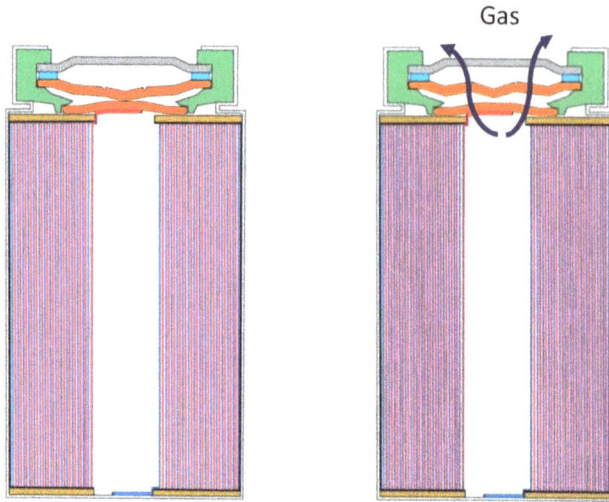

Figure 5.6 Current interruptible device being activated. Mainly due to gas generation caused by overvoltage, the diaphragm inside the battery cap deforms and shuts off the current path. As explained later, this is an effective device for preventing overcharging, which is considered the most dangerous for lithium-ion batteries. Thanks to my LinkedIn friend Dr Sylvester (Avijit) Gomess for the illustration.

a mechanism that draws current through a diaphragm, and, when the battery is in a dangerous state of overcharging, the pressure of the gas generated inside causes the diaphragm to deform, shutting off the current (Figure 5.6). To ensure the operation of the current interruptible device (CID), the electrolyte contains an additive that decomposes and becomes a gas at a certain potential. As described above, the 18650-type cylindrical lithium-ion battery has been developed over many years, and because of its long history, it can be said to be a fairly well-developed lithium-ion battery in terms of safety. The all-tab cylindrical lithium-ion battery was developed by various companies at the beginning of the 21st century. The ideal current-collecting structure of the all-tab battery provided excellent power output and long life, as expected. However, the treatment of the cathode terminals did not go well, resulting in many smoking accidents due to contact between the anode and the cathode tabs.

Of the short circuits between the electrodes, the heat generated by contact between the negative electrode active material and the aluminium current collector foil is the greatest. In the past, there have been many incidents of fire caused by this. All areas where the aluminium foil is exposed need to be insulated with Kapton tape or similar. Beginners who are going to make batteries should carefully examine the components of batteries by disassembling those that have a proven track record.

5.3   Square (Prismatic) Type

In addition to the cylindrical type, there are two other types of lithium-ion battery casings: the square type, which uses a square metal can, and the laminated type, which uses a soft laminated pouch. The square and pouch types are available in two types of inner packaging: the flat-wound type, in which the battery is rolled flat, and the laminated type, in which the separator and electrodes are stacked one on top of the other. The flat-wound element is a variant of the roll-wound type and was widely used for small portable applications. The flat-wound element is also used for BEV applications, but the rounded edges of the element make it less volumetrically efficient for square cases. For BEV applications, where volumetric energy density is important, stacked types are increasingly being used to fill the case to its full capacity. Since flat-wound elements are derived from cylinders, they have a long history and are well established, which means that they can be manufactured quickly and will continue to be used for medium-sized square cells in the future. In the past, winding was inherently faster than stacking one layer at a time. However, BEV cells are becoming larger and larger, and the large projected area (called the footprint area) of such large cells means that the number of layers can be reduced, relatively. For large cells, the stacking type with good lifetime characteristics and volume efficiency is likely to become more common. The most difficult part of stacking is the positioning of the separator. It is quite difficult to position the soft and fluffy separator. Recently, separators have been applied with a special adhesive and integrated with the anode (or cathode) in advance, and then assembled at high speed in the subsequent processes. Stacking using this method is comparable to the speed of forming conventional cylindrical elements, or even faster in terms of the area stacked per unit time. To maintain the exact distance between the cathode and anode (distance between poles), a certain amount of pressure is necessary. In the cylindrical type, the tension at the time of winding can determine the distance between the poles. However, in lamination and flat winding, it is necessary to apply some pressure from the outside. In the case of flat winding, the end faces are folded, so it is difficult to determine the distance between the poles, and deterioration is likely to progress from that part. Basically, stacked cells have better durability and output. In cylindrical cells, the case functions as a negative terminal, while in square cells, the positive and negative terminals are often attached to the cap. I have seen some large cells with a negative case, but I thought this was a somewhat scary situation because the huge negative terminal was exposed. I have not seen many of these recently, however. Whether flat-wound or stacked, the output can be larger than for cylindrical cells because the current is drawn through a thicker terminal.

The dried element is placed in a square case, the negative and positive terminals are attached to a cap, and the case and cap are sealed by laser welding. The cap has a hole in it through which electrolyte is injected. As with cylinders, vacuum and pressure are repeatedly applied to infiltrate the electrolyte. After sealing the electrolyte inlet using laser welding, initialization and aging are performed. After the initialization gas is removed and the inlet is sealed.

The distance between the poles cannot be accurately determined by the winding tension of the separator, as is the case with cylindrical cells, so it is necessary to apply the appropriate pressure at the module or pack stage after forming the cell.

In terms of homogeneity, the stacked type of cell is currently the best, and it is easy to obtain cells with good lifetime characteristics. All-tab cylindrical batteries, which excel in terms of degradation resistance and lifespan, are adopted by major electric vehicle manufacturers.

5.3.1 Treatment of the Cathode Terminal

Here I will repeat myself, but it is important. The uncoated portions of the cathode and anode are very delicate. The coated part is coated with active material, and therefore, the resistance is higher than that of the bare metal part. In general, it is known that the amount of heat generated is the highest when the cathode aluminium part contacts the anode active material, rather than when the cathode and anode active materials contact each other. In any case, the exposed metal part of the electrode should be strictly protected with polyimide tape or other means. It is believed that a large percentage of smoke and fire accidents caused by poor design in the past occurred when the aluminium portion of the cathode came into contact with the anode compound for some reason.

5.4   Pouch Type

Although it may sound a bit rough and ready, the pouch type cell can be described as a retort inside a pouch that replaces a square outer case. Inside the pouch is an element formed by a flat-rolled or laminated element. Although the production process is similar to that of the square type, it is not as robust, and the pouch expands as soon as gas is generated inside. Incidentally, CIDs can also be built into square cells, but I have never seen such a structure in a pouch cell. But how about these days? As I will discuss later, the biggest fear in batteries is overcharging. For this reason, it is very important that the battery itself has a CID. Since it is difficult to build in a CID for pouches, I think it is necessary to have a strict overcharge protection mechanism, such as a battery management system (BMS). After the electrolyte is poured, the cell is temporarily sealed and the reaction gases from initialization and aging are removed before the cell is evacuated and sealed again. Compared to rectangular cells, pouch cells offer greater design flexibility for internal electrodes, making it easier to design larger lead terminals and thus more readily increase output. Many recent EV batteries do not include a CID. This is likely because a CID provides resistance to current. In such cells, the BMS must implement control to prevent overcharging. The initial capital investment is relatively low, the shape is highly flexible, and it is easy to obtain a thin structure with a large surface area, which facilitates cooling of the heat generated when the battery is in use. As with the rectangular type cells, the distance between the poles cannot be precisely determined by the tension of the separator winding, as is the case for the cylindrical type, so it is necessary to apply appropriate pressure to determine the distance between the poles when the cells are melded into modules or packs. It is necessary to understand that a large pouch (and a square cell) for BEVs is a semi-finished battery, and it is only when it is combined with a pressurization mechanism when it is made into a pack, *etc.*, that it can demonstrate its true performance. How long should the pouch seal be? This is quite difficult to determine. Depending on the guaranteed age of the cell, it will be necessary to determine the seal width by leaving samples of product with different seal widths in a harsh environment and measuring the amount of moisture ingress. In the past, it was said to be 1 year per mm, but I am not so sure now.

Figure 5.7 shows an assembly flow diagram of a pouch-type lithium-ion battery. Since the laminate pouch is not a pressure-resistant container, the gases generated during initialization must be removed

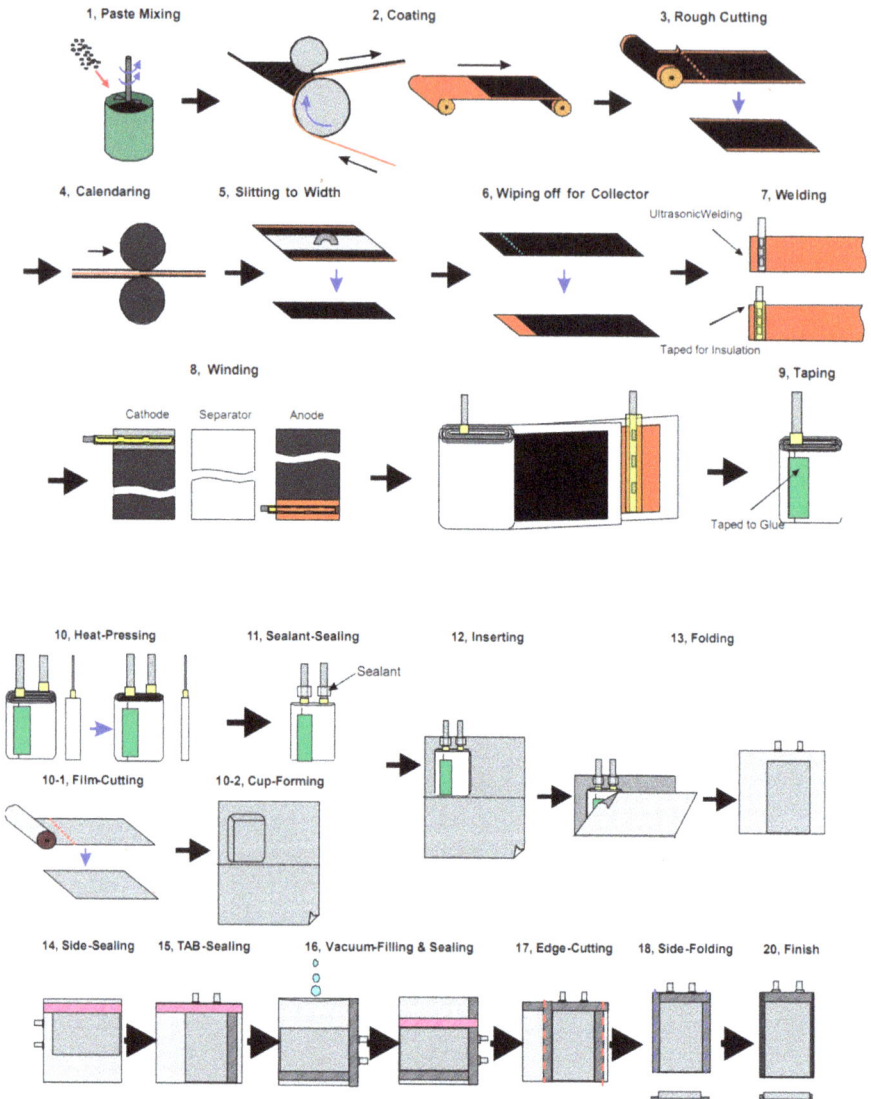

Figure 5.7 Assembly flow diagram of pouch-type lithium-ion battery. Illustration courtesy of Hohsen.

in advance. The area in which the output and material properties of the electrode can be measured most accurately is actually in the laminate. A single-piece cell with only one cathode and one anode facing each other is also a development process that comes next to measurement with coin cells. There is also expertise in shallow drawing of the laminate, but this is not discussed here.

Lithium-ion batteries are, to begin with, averse to moisture. There-fore, the idea of using aluminium laminate pouches to form the cells is something that anyone would come up with. About 30 years ago, I tried to file a patent for using laminate as an outer packaging material, but the same idea existed for the lead-acid battery, and it was neces-sary to narrow down the rights significantly. The inside of the lami-nate film should be a polypropylene-based resin that is resistant to electrolyte. Nylon or polyethylene terephthalate (PET) is used on the outside, but the inside often has a polypropylene liner. The adhesion between the aluminium foil and the laminate resin also requires skill. The aluminium surface is degreased and treated with anodized alu-minium, aluminium hydroxide treatment on the surface to roughen it up in advance, and then an adhesive that is resistant to electrolytic solution is used to bond them together. It is a funny story for me now, but once, when I ordered food-grade laminate to make a prototype battery, in order to reduce the cost of batteries, the resin used to bond the aluminium foil and polypropylene dissolved in the electrolyte, leaving only a transparent bag and a visible cell. At that time, I was in a hurry. The food-grade laminate was laminated with urethane, and electrolyte had penetrated through the pinholes on the inside, dissolving the adhesive. Drawing and molding of the laminate also requires a certain technique. There are also combinations of resins to improve the moldability of laminates. The welding part of the drawer terminal to the laminate film also requires ingenuity. The aluminium of the positive terminal surface must be treated with a hydrating film or a thin phosphoric acid film, and then a sealant made of polypropyl-ene must be applied in advance. In this case, modified polypropylene is used as the adhesive, which has polarity to improve adhesion. Since the surface of aluminium is subject to change, it does not adhere well to resin, and I remember that it was very difficult to develop this technology.

5.5 Electrolyte Injection, Infiltration, Initialization, and Aging

Electrolyte injection, impregnation, initialization, and aging are the most important and know-how-intensive post-processing steps after electrodes. Voltage must be applied only after the electrolyte has been thoroughly impregnated and the electrolyte has spread through every part of the separator and both electrodes. If voltage is applied without the electrolyte being filled up, electrode utilization and SEI formation on the negative electrode will be uneven, resulting in unexpected deterioration of the electrodes. A more advanced technique for filling electrolyte is to apply a voltage to the terminals in the range below the voltage at which the SEI is formed, which will increase the electrolyte impregnation. The purpose of initialization is to form a good-quality film on the surface of the anode. The performance of the film formed changes depending on the electrolyte solvent and additives used, the voltage applied, and the temperature of the battery, and this has a significant impact on the subsequent life of the battery. This is where the know-how of each manufacturing company is concentrated. Since initialization generates a large amount of gas, square and pouch batteries are resealed after initialization (or aging, as the case may be). Cylindrical types are basically sealed at the start. Aging is performed after initialization by removing the power supply and inducing self-discharge of the cells. To put it in a simple way, aging is a process to stabilize the SEI film created by initialization and to deteriorate the parts that are prone to deterioration in advance, in the factory. The manufacturer must guarantee the life of the cell to some extent. Degradation tends to be greater at the beginning and gradually decreases in severity. Therefore, if the cells are degraded from the beginning, the degradation rate after shipment can be reduced (made visible). On the other hand, if the product is deteriorated in the factory in advance, the product retention time becomes longer, which is disadvantageous from a manufacturing cost perspective. Aging time also serves as a short-circuit check process for products, since cells with internal shorts can be detected and removed. The temperature is a very important factor for initialization and aging. In the past, when I performed impregnation and initial aging in a very cold factory, the electrolyte deteriorated so severely that it turned an abnormal color. The cause was that when the electrolyte was poured in the cold, the viscosity of the electrolyte was so high that it did not spread to every corner of the electrode, and the voltage was applied at that stage, resulting in partial overvoltage. The initialization and aging processes are the first time voltage is applied to a battery, and, in the worst-case scenario, if there is a short circuit, the battery will burn. In fact, when there is a fire accident in a battery factory, the initialization and aging processes should be the first suspects. At a factory I was involved in, the aging process was conducted in a "separate" building so that the main part of the factory would not be damaged in case of an emergency. Initialization and aging can take several days to several weeks, which increases the cost of battery production. Some manufacturers who can produce their own batteries and BEVs at the same time have

Figure 5.8   A photograph of the degraded electrode surface of a flat-rolled cell.

the idea of pouring the fluid into the battery and immediately install-ing it in the vehicle, so that the initialization and aging process can be done in the vehicle. I am not sure if this is possible, but it is the ultimate solution.

Again, the processes that require the most know-how after electrode formation are electrolyte impregnation, initialization, and aging. Even if initialization and aging are carried out properly, there are cases where deterioration progresses due to factors such as the shape of the cell. Figure 5.8 shows a photograph of the negative electrode of a flat-rolled cell that has deteriorated. You can see the change in color of the electrode, which has been bent and wrinkled. This is a part of the electrode that has been subjected to some kind of stress compared to other parts. In order to improve the life characteristics, it is neces-sary to make efforts to reduce such uneven parts in both design and manufacturing.

5.6   Cell Inspection Items

Capacitance, self-discharge characteristics, and internal resistance are measured in order to inspect the quality of the cell. Any defects in appearance must also be checked. Resistance is measured by direct current internal resistance (DCIR) or alternating current (AC). Alter-nating current resistance (ACR) is only a guide, and rechargeable bat-teries used with direct current should be guaranteed by DCIR. If the OCV after aging is checked and is out of range, the battery should be excluded as having abnormal self-discharge.

6 Safety and Degradation Mechanisms of Lithium-ion Batteries: How the Battery Management System Works

6.1 Thermal Runaway of Lithium-ion Batteries

This chapter explains thermal runaway in lithium-ion batteries. The range of applications for lithium-ion batteries is expanding, and they are also being used in large devices, such as battery electric vehicles. Lithium-ion battery fires are one of the most important safety issues. The various factors involved in thermal runaway in lithium-ion batteries are the subject of the latest academic research, but there is no established theory. In this chapter, I put forward an explanation based on my own observations.

Although lithium-ion batteries have excellent characteristics, they are plagued by thermal runaway, which can induce smoking and ignition due to the high energy content and the use of flammable electrolyte inside the battery. In recent years, as batteries have become larger, the scale of accidents has also increased. At the beginning of thermal runaway, white smoke will occur (Figure 6.1a). The white smoke is heavy, spreading as if crawling along the ground. The composition of this white smoke is a mixture of various substances. Under non-reactive

RSC Foundations No. 4
Li-ion Batteries
By Hitoshi Nakamura
© Hitoshi Nakamura 2025
Published by the Royal Society of Chemistry, www.rsc.org

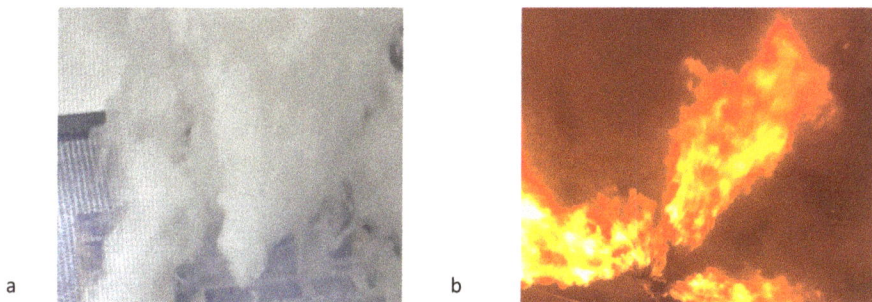

Figure 6.1   When the cell is heated, the first white smoke is generated (a). A large flame rises as highly heated parts ignite the white smoke (b) (photo by the author).

conditions, it is thought to include combustible hydrocarbon gases, carbon monoxide, hydrogen gas, and other gases generated when surrounding combustible materials are partially oxidized by oxygen generated at the cathode. It also contains gases produced from mist in the electrolyte.[3] The white smoke spreads and mixes with external oxygen and ignites when it comes in contact with red-hot parts, such as sparks or short circuits, and turns into flames. In the absence of sufficient oxygen (hypoxic conditions), the white smoke may be oxidized and turn into black smoke (Figure 6.1b). When the black smoke is mixed with oxygen that came from outside and a high-temperature heat source is present, it also ignites and burns violently. I will repeat this more succinctly: in the case of thermal runaway of lithium-ion batteries, white smoke is the first characteristic to be generated, and then the battery is exposed to a high-temperature heat source and burns explosively. This reaction is accompanied by the generation of black smoke. The black smoke itself ignites and burns when there is an ignition source in an atmosphere with sufficient oxygen, causing extensive damage.

6.2   Stories About Thermal Runaway and Personal Definition of Thermal Runaway

What is the state of thermal runaway? There are various definitions, but my understanding is that it is a state in which the oxygen released by the decomposition of the cathode continuously oxidizes (or burns) the surrounding combustible materials (electrolyte, separator, aluminium, *etc.*). When thermal runaway occurs, the cathode disintegrates continuously, and the combustion reaction by means of the release of oxygen progresses, resulting in an uncontrollable state.[1,2]

In a lithium-ion battery, the electrolyte, as the fuel, and the cathode material made of metal oxide, as the oxygen source, coexist, and so combustion occurs even without an external oxygen supply. This is similar to a runaway rocket engine. If the outer case is damaged by an explosion or other accidental damage caused by thermal runaway, oxygen can flow in from the surroundings, further increasing the intensity of the fire. If you happen to be on the scene of a battery fire, you should consider evacuating the area because thermal runaway of lithium-ion batteries releases large amounts of combustible carbon monoxide, hydrogen, and hydrocarbon gases.[3] Also, flammable gases may reignite, so backdrafts should be prepared for when ventilating the fire scene. Once a battery has entered a state of thermal runaway, it cannot be easily extinguished. Even if the supply of external oxygen is shut off, using carbon dioxide as in the case of a normal fire, the effect may be limited because the battery will burn with oxygen supplied from inside. However, shutting off the oxygen can prevent secondary combustion and prevent the damage from spreading. The most effective way to extinguish the fire is to pour large amounts of water or sand onto the fire to lower the temperature and stop the thermochemical reaction.

6.3   Voltage or Temperature?

What are the most dangerous modes of lithium-ion batteries? If thermal runaway is the continuous release of oxygen due to the collapse of cathode oxide crystals, then heating and overcharging are the most dangerous situations. Overcharging, in particular, is accompanied by heat generation due to decomposition of electrolyte and other substances, and so heating and overcharging occur at the same time and are very dangerous. The cathode is a compound formed by the neutralization of oxides with lithium. As the lithium-ion battery is charged, lithium ions move from the cathode to the anode. In the cathode without lithium ions, the metal oxide exhibits its original strong oxidizing property and easily reacts with the surrounding combustibles. When lithium ions are removed, oxygen in the crystal is next affected by electron withdrawal, and extremely active oxygen radicals with no electrons are released from the crystal. If it is indeed oxygen radicals that are released, then these are one of the most dangerous substances on Earth. They have the property of withdrawing electrons from anything they come into contact with, and immediately oxidizing it. When this happens, the surrounding combustible materials are oxidized, or burned, without a second thought.

The four major hazardous modes of batteries described in the following section are all triggers that lead to thermal runaway. However, considering the above description, the most dangerous trigger is unintentional overcharging of the battery, followed by an internal short circuit that occurs near full charge. In order to ensure system safety, the battery management system (BMS) must at least be able to detect overcharging and abnormal temperature increases. It would be even better if the BMS could detect internal short circuits in the battery, but this is quite difficult technically.

6.4 Dangerous Battery Modes

As explained, a lithium-ion battery will undergo irreversible thermal runaway when it reaches a temperature or cathode potential at which the cathode decomposes. For example, the temperature at which the cathode decomposes and releases oxygen is 200 °C or higher for ternary cathode materials, and 4.6 V or higher for voltage. In other words, it is necessary to devise ways to prevent the battery from reaching such a state.

To remove the cause or shut down the system before the cathode temperature reaches 200 °C or 4.6 V or higher, it is necessary to add a device on the cell side, such as a current interrupt device (CID), or on the circuit side, such as a BMS. It is also theoretically possible to prevent thermal runaway reactions in advance by detecting cells that are about to reach thermal runaway early and actively cooling such cells or removing voltage from them. However, the reality is that there are limited methods to identify cells that have reached a dangerous state and to apply cooling or rapid discharge.

In my experience, thermal runaway does not occur unless several conditions are combined, such as overcharging and overheating, and, in many cases, thermal runaway does not occur if the temperature is lowered by slightly blowing air over the cell, for example.

It is important to detect a cell that has reached a dangerous mode as soon as possible, and then the cell should be rapidly discharged or cooled down.

6.4.1 Overcharge

In batteries using ternary cathode materials, which are typical cathode materials for lithium-ion batteries, when the voltage exceeds 4.6 V, the materials start to decompose and release oxygen from the inside of the crystals. At that point, the battery is assumed to be in a state of storing excess energy of about 150–200% of normal.

Since the energy stored before thermal runaway is greater in overcharging, it is assumed that the combustion power will be stronger and cause the most damage when thermal runaway occurs in overcharging mode. System designers need to protect batteries from overcharging with at least double protection circuits to prevent overcharging. As mentioned earlier, some cylindrical and square cells are equipped with a kind of pressure-operated fuse, called a CID, to shut off overcharge. However, it is difficult to install a CID in a pouch cell due to its structure.

Since the basic idea of a fail-safe is double protection, it is necessary to consider some kind of double protection on the BMS circuit side for overcharge protection for pouch cells.

The overcharge condition can be detected from the battery charge waveform. To compensate for overcharging, lithium is removed from the cathode, and when the lithium ions inside the crystal are gone the oxygen in the crystal begins to be pulled out. In this case, especially in a layered metal oxide system, a unique voltage fluctuation may arise. The cell temperature begins to rise as soon as the unique voltage fluctuations associated with the onset of oxygen extraction are experienced. This is one signal that can detect thermal runaway.

To understand thermal runaway, it is recommended to observe the overcharging of lithium-ion batteries with layered metal oxide system cathodes. Of course, this should only be done after ensuring safety procedures are in place.

When an overcharge is initiated, the voltage of the lithium-ion battery continues to rise at regular intervals for a while. The temperature rises slightly. Eventually, when the voltage begins to exceed 4.6 V, a situation arises in which it becomes difficult to raise the voltage. It is thought that the lithium ions in the cathode have been depleted. Next, the voltage begins to drop gradually. It is thought that lithium ions are depleted and oxygen inside the crystal begins to be drawn out, and the surrounding electrolyte begins to decompose. The temperature begins to rise rapidly, triggered by the drop in voltage.

As the battery temperature rises, the safety valve may open due to the pressure increase caused by the evaporated electrolyte and decomposition gases, depending on the type of battery. When the safety valve opens, the battery voltage begins to rise rapidly. There are two possible causes: one is the resistance increase due to separator shutdown, and the other is the resistance increase due to evaporation and depletion of the electrolyte. When the cell temperature exceeds 140–160 °C, the separator melts, inducing a short circuit and causing the battery temperature to rise quickly, leading to thermal runaway. The heat sensor can only be attached to the outside of the battery cell. Although thermal runaway often occurs at temperatures not exceeding 140–160 °C on the outside of the battery, the actual internal temperature of the battery cell may well exceed 140–160 °C.

The figure of 140–160 °C is for cases where a separator without a ceramic coating is used. In fact, for cells that use a separator with a ceramic coating, the temperature at which runaway begins is increased by around 10–20 °C. However, the user of the cell will not know this. Therefore, if the temperature exceeds 140–160 °C, the separator is considered to have begun melting and is in an extremely dangerous state.

6.4.2 Internal Short Circuit

An internal short circuit refers to direct contact between the positive and negative electrodes as a result of damage to the separator. Although the voltage is locally lowered, a large current flows and heat is generated rapidly, which often exceeds the temperature at which the cathode begins to decompose, resulting in thermal runaway.

The danger of heat generation during thermal runaway depends to some extent on the state of charge (SOC) of the battery, and, in particular, the possibility of thermal runaway increases when the voltage is high and energy is high, such as when the battery is fully charged, so care must be taken.

In the case of an internal short circuit, current flows into the shorted area from the surrounding area, causing the battery to reach a high temperature in a relatively short period of time. This is understood to cause the cathode material in a high SOC state to be exposed to high temperatures and eventually to undergo crystal collapse with oxygen release, and so leading to thermal runaway.

There are internal and external factors that cause an internal short circuit. External factors include cases where the separator breaks due to deformation caused by a large external force being exerted on the cell, which results in a short circuit. Endogenous factors include those that occur in a low-temperature environment and internal factors that are incorporated at the time of manufacture. A short circuit caused by dendrites, which occurs during repeated charging at low temperatures, is an internal factor that can be solved by limiting regeneration and charging at low temperatures through a BMS or other means.

In some cases, metal foreign matter introduced during the manufacturing process can grow abnormally and cause short circuits. In such instances, danger may be averted through the BMS or vigilance of the user. Manufacturers must take particular care to prevent metal foreign matter from entering the cell interior. Users employing the cells should take measures such as purchasing batteries from established and reliable cell manufacturers. In some cases, it may be necessary to inspect the manufacturer's factory.

6.4.3 External Short Circuit

An external short circuit is a direct short circuit between the positive and negative electrode terminals of a cell. The large amount of heat generated induces melting of the internal separator, resulting in an internal short circuit and, in the worst case, thermal runaway. In addition, since a huge current flows through the short circuit, the metal parts that are short-circuited rapidly become red-hot, and this heat may also induce thermal runaway.

6.4.4 Searing Heat (Heating)

When a battery is exposed to high temperatures, the internal separator melts, leading to a short circuit and thermal runaway. When a battery is exposed to external heat, it may begin to generate heat on its own, called self-heating. Since self-heating is associated with disintegration and reconstitution of the solid–electrolyte interface (SEI) on the anode surface, which may start at around 60 °C, it is necessary to control the temperature of the battery, especially when charging, by not starting the charging process until the battery temperature falls below the temperature specified by the manufacturer. Thermal runaway due to searing heat can also occur during incorrect use, such as when a cell phone is accidentally placed on a stove top. If the battery is close to being fully charged and the cell temperature exceeds 140–160 °C, a short circuit due to the melting of the separator caused by the additional external heat may result in thermal runaway. In addition, cathode materials in a high SOC state are at risk of continuous thermal runaway due to oxygen coming from inside the cathode when the cell temperature exceeds 200 °C.

The so-called "pack thermal runaway" initially causes a specific cell to experience runaway. Next, the heat spreads to other batteries, causing them to explode. In countries with strict standards, a battery must pass a test to confirm that it has sufficient functions to prevent such a thermal chain reaction, called a thermal chain test. Thermal runaway of a specific cell is often caused by an internal short circuit. On the other hand, anti-thermal chaining requires thermal insulation between cells or resistance to high cell temperature.

6.4.5 Internal Shorts Due to Manufacturing Factors That Cannot Be Controlled by the Battery User

As I mentioned in Section 6.4.2, there are two sources of internal short circuits in battery cells. One is the destruction of the cell due to a crash, *etc.*, and the other is a factor contained within the cell. The factors that are contained in the cell can be further divided into two types: those caused by rapid charging in a low-temperature environment and those caused by foreign matter that is introduced into the cell during the manufacturing process. The manufacturing process for lithium-ion batteries, which involves a wide variety of processes, requires skilful control of foreign substances that could lead to internal shorts, and accidents caused by internal shorts created in the manufacturing process have been seen in the past, even in cells manufactured by leading manufacturers. In recent years, a number of accidents are thought to have been caused by design factors, such as unreasonable processing of the positive electrode, distortion of the cell due to vibration (for example), and contact with the negative electrode, all of which may lead to ignition.

In the early 2000s, a PC equipped with a battery generated heat during a meeting and burned out. The accident was widely reported because it occurred in the public eye. In parallel with this incident, a major complaint was made against other companies' batteries as well, claiming that abnormal heat generation was observed.[4]

In the case of the computer, there was a strong possibility that the lithium-ion battery installed in the computer had ignited due to an internal short circuit, resulting in a fire and smoke accident. In response to this incident, a method for evaluating the internal short-circuit resistance of lithium-ion batteries was devised, standardized, and established, known as the forced internal short circuit (FISC) test for lithium-ion batteries. The method is straightforward:

the battery is disassembled and reassembled with a small nickel L-shaped piece placed between the cathode and separator. While the FISC test is direct and easy to understand, it is a complex method that involves disassembling the battery, inserting the small piece of nickel, and then pressurizing it, making it highly dependent on the skill of the operator and making it difficult to reproduce the results obtained. To solve this problem, a method has been proposed in which a nail is mechanically inserted from the outside of the battery case to induce a small internal short circuit; this is not affected by the operator's skill (direct internal short circuit, DISC).

The nails must be stopped by a short circuit in one layer (one set of separator–anode–separator–cathode), which requires fairly precise control of the nails. Despite the use of precision machinery, there is still a problem with reproducibility, and the system has not yet become widely used. Standardization is also lagging.

In many cases, defects inside the cell that could lead to an internal short circuit due to manufacturing or usage conditions affect the temperature at which the cell begins to generate internal heat. Therefore, the use of an accelerating rate calorimeter (ARC), a device that can precisely detect the self-heating of a cell, is increasingly being applied (Figure 6.2).

However, there is still no method available to accurately evaluate the internal short-circuit resistance of cells. It is said that the majority of vehicle fires caused by cells are due to internal short circuits, and so it is expected that a method to accurately determine the internal short-circuit resistance of manufactured cells will be developed.

ARC can be used to accurately capture the heat generation behavior of cells,[5,6] which is useful for pack and BMS design.

6.5 How the BMS Works

One of the functions of the BMS is to control the safe operation of the battery.

The BMS must be designed to detect and eliminate the four major cell factors for non-safe cells discussed above. Overcharging is the most serious hazard and requires a double fail-safe combined with the cells. Cells for which it is difficult to provide voltage shutdown capability in the event of overcharge, such as laminated pouches, require double overcharge protection on the BMS side. Overcharging is a situation that must be avoided at all costs, so overcharging protection should be implemented for all cells. Next, the BMS needs to monitor

Temperature as a Function of Time

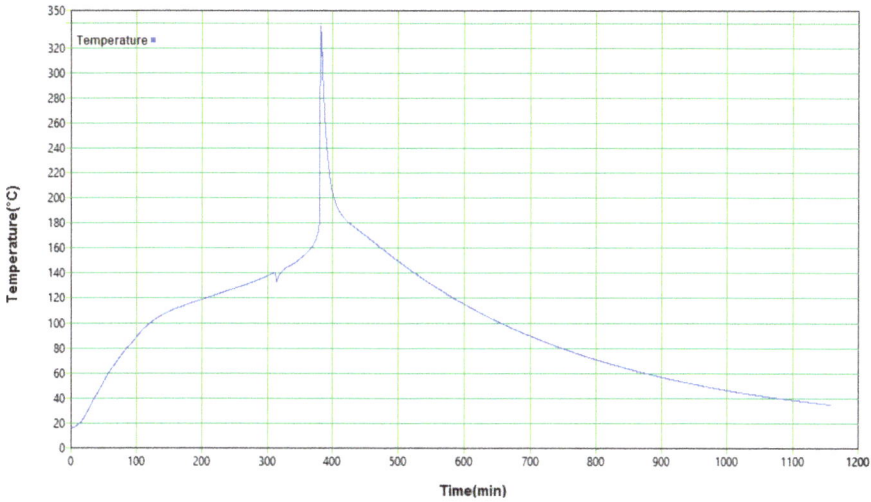

Figure 6.2 Example of ARC measurements for 18650-type cells (data collected by the author). The horizontal axis shows time, and the vertical axis shows the temperature reached by the cell. The difference between DSC (differential scanning calorimetry) is that ARC enters a mode to check for the presence of self-heating in the cell when the temperature is raised, and, if the cell's heating is not confirmed, it enters the next temperature rise step.

heat to detect abnormal cell heating or abnormal low temperatures, and to stop discharge or interrupt charging. As mentioned earlier, if the system can actively detect dangerous states in the cells and actively cool them down, it is possible to avoid thermal runaway. Even if there is no room to install such a cooling system, it is desirable to monitor the temperature of all cells with a BMS, although this is expensive. If one cell exceeds its limit, measures such as shutting down the entire system can be taken. In the case of a mobile vehicle, such as a BEV, sudden shutdown of the battery system is not desirable, so the BMS needs to coordinate with the main system by issuing warnings before the shutdown occurs. Controller area networks (CANs) and other communication standards are used for such communication.

The BMS may be equipped with a simple remaining capacity estimation circuit. The remaining capacity estimation can be done using one of the following methods: calculating the remaining capacity from the difference between the integrated current during charging (called Coulomb counting) and the integrated current during discharge, measuring and recording the relationship between the SOC and voltage in advance, and calculating the SOC by comparing it with the voltage at no load (OCV, open circuit voltage). A combination of these methods is also available. A circuit that can measure the residual life of cells may also be provided. A simple method is to compare the integrated value of the charging current and estimate the life based on the degree of change, while an advanced system may be installed to estimate the remaining capacity based on the waveform during charging and changes in cell resistance. Methods for estimating battery life will be described in detail in Chapter 7.

A balancer to balance the cells may also be built in. In my opinion, cell variation should be managed by quality control in the manufacturing process, and balancers should not be used, if at all possible. In fact, in a well-made pack, the capacity variation of the battery is often matched to the minimum limit in advance, so a balancer is rarely needed, only when the cell combination is very bad. If the cell variation is too large, the life of the pack cannot be guaranteed.

References

1. M. Hirooka, T. Sekiya, Y. Omomo, M. Yamada, H. Katayama, T. Okumura, Y. Yamada and K. Ariyoshi, Degradation mechanism of $LiCoO_2$ under float charge conditions and high temperatures, *Electrochim. Acta*, 2019, **320**, 134596.
2. Y. Dai and A. Panahi, Thermal runaway process in lithium-ion batteries: A review, *Next Energy*, 2025, **6**, 100186.
3. T. Hayashi, J. Okada, E. Toda, R. Kuzuo, N. Oshimura, N. Kuwata and J. Kawamura, Degradation mechanism of $LiNi_{0.82}Co_{0.15}Al_{0.03}O_2$ positive electrodes of a lithium-ion battery by a long-term cycling test, *J. Electrochem. Soc.*, 2014, **161**(6), A1007.
4. P. J. Bugryniec, E. G. Resendiz, S. M. Nwophoke, S. Khanna, C. James and S. F. Brown, Review of gas emissions from lithium-ion battery thermal runaway failure — Considering toxic and flammable compounds, *J. Energy Storage*, 2024, **87**, 111288.
5. X. Feng, X. He, M. Ouyang, L. Wang, L. Lu, D. Ren and S. Santhanagopalan, A Coupled Electrochemical-Thermal Failure Model for Predicting the Thermal Runaway Behavior of Lithium-Ion Batteries, *J. Electrochem. Soc.*, 2018, **165**, A3748–A3765.
6. T. Waldmann and M. Wohlfahrt-Mehrens, Effects of rest time after Li plating on safety behavior—ARC tests with commercial high-energy 18650 Li-ion cells, *Electrochim. Acta*, 2017, **230**, 454–460.

7 State of Health of Lithium-ion Batteries

In this chapter, I explain the technology for analyzing and predicting the state of health of lithium-ion batteries. The technology for predicting the lifespan of lithium-ion batteries is a relatively new research field that is still in the process of development. To understand this chapter, you will need to have understood all the previous chapters. Battery life analysis is a very advanced field of research, and you need to read a lot of papers to get a bird's-eye view of the state of development. I have tried to cover as much as possible of this field, and have provided an overview of the main analysis methods. There is much more to consider, so please continue your study using the references.

7.1 Battery Degradation Modes

Battery degradation involves all aspects of battery technology. Here, I'll briefly and frankly discuss the factors causing battery degradation, the resulting degradation analysis, and perspectives on extending battery life.

As I have written many times in the past, in the range of oxidation resistance or reduction resistance of electrolytes (called the potential window), the reduction resistance of the non-protonic solvents currently in wide use is not sufficient, and most electrolyte solvents are

RSC Foundations No. 4
Li-ion Batteries
By Hitoshi Nakamura
© Hitoshi Nakamura 2025
Published by the Royal Society of Chemistry, www.rsc.org

will thermodynamically reduce and decompose at the potential at which the lithium ions insert and extract graphite.[1] However, nature is good, and the decomposition products of the electrolyte are deposited on the electrode surface, like a film, which becomes a protective film and suppresses further decomposition of electrolyte solvents. This allows the battery to function, even though it is used in a potential range where reductive decomposition would normally take place. In addition, additives that prevent the decomposition of the solvent by itself, before the solvent decomposes by reduction, have become widely used, enabling stable operation over hundreds of cycles.[1–3]

However, just because it takes a long time for decomposition to occur does not mean that the intrinsic reductive decomposition reaction has disappeared. Of course, the factors that contribute to degradation are complex. However, it is easy to understand that the degradation of a properly designed lithium-ion battery begins at the graphite surface of the negative electrode.[4,5]

A lithium-ion battery is a battery that compensates for the transfer of electrons (charge) by the transfer of lithium ions. Therefore, when the concentration of lithium ions available for operation in the system decreases, the battery deteriorates because it is unable to transport charge.

The expansion and contraction of graphite in the negative electrode due to charging and discharging is a well-known phenomenon. When the bulk graphite expands and contracts, the protective coating formed on the graphite surface cracks, and the electrolyte reductively decomposes in the cracked area to repair the film. The film also takes in more lithium ions, which are immobilized. Each time the film is re-formed, it traps lithium ions, and so the number of freely moving lithium ions (*i.e.* those that can contribute to charge transport) decreases.

If the number of freely moving lithium ions decreases, then the number of lithium ions that can return to the cathode during charging decreases. If insufficient lithium ions are returned to the cathode, the potential of the cathode cannot be fully restored. This causes the cathode to deviate from its original capacity range and operate in a range in which the cathode utilization rate is lower than the original design. This is called capacity misalignment.[6,7]

The protective film (solid–electrolyte interface (SEI)) also serves to protect the anode surface by inhibiting reactions, but, conversely, it is a resistive body through which it is difficult for lithium ions to pass. The thicker the protective film, the greater the resistance of the battery, and the drop in IR increases, making it impossible to obtain the maximum capacity.

Thus, in a lithium-ion battery, deterioration of the negative electrode is the starting point for a reduction in the overall capacity of the battery. So, is it sufficient to suppress side reactions between the negative electrode and electrolyte to reduce the degradation of lithium-ion batteries? First of all, it is necessary to form a homogeneous anode, carefully initialize the anode, and obtain an appropriate combination of additives and materials. If you do these things steadily, you will be amazed at the improvement in lifespan. As I have mentioned before, the higher the crystallinity of the graphite anode, the easier it is to decompose the electrolyte solvent. Therefore, mixing artificial graphite with low-crystallinity graphite, or coating the surface of graphite with amorphous carbon, will greatly improve the life of the anode.

It is also appropriate to ask whether forming electrodes using amorphous carbon extends the life of the electrodes. However, since amorphous carbon has a low initial charge–discharge efficiency, it is necessary to provide extra lithium ions during initialization. For example, the electrolyte concentration can be increased, the percentage of positive electrode can be increased, or metallic lithium can be attached to the negative electrode.

There are various methods to estimate the degradation behavior of batteries. The protective coating known as the SEI deteriorates and thickens over time. Linking this change to capacitance offers a useful non-destructive method for estimating degradation.

There are two major non-destructive methods for analyzing the remaining capacity of a battery: the impedance method, to measure resistance, and comparing the waveform of the charging and discharging sides.[8–10] Since both methods are highly susceptible to temperature variations, measurements must be conducted under constant temperature conditions.

As battery degradation progresses, battery safety also changes, and detailed observation by accelerating rate calorimetry (ARC) shows that the self-heating temperature tends to decrease as battery degradation progresses. However, the maximum calories generated by the battery become lower, and the battery becomes more stable and safer. Thus, overall the batteries become more unstable, but the ongoing damage tends to be smaller in many cases. This seems to make sense if it is considered that the electrolyte is consumed as the amount of the thermodynamically unstable protective film increases, so the amount of electrolyte used as fuel decreases.

Since electrolyte is being consumed during degradation, there is a tendency to assume that stocking a large amount of electrolyte in advance will have a positive effect on battery life. In fact, as mentioned before, the life of 18650-type batteries tends to shorten as the energy density increases, and is inversely proportional to the amount of electrolyte that can be stored in the battery. It is also known that lithium-ion batteries develop an abnormal protective coating when charged at low temperatures. The lower the temperature, the stronger the interaction between the solvated lithium ion and solvent becomes, and the harder it is for the lithium ion and solvent to be separated. When a lithium ion tries to dive into the gap between carbon, the solvent is usually removed, but the lithium ion tries to take the solvent with it, which destroys the carbon and decomposes the solvent. This causes abnormal development of the protective film, which is dangerous and can sometimes break the separator, causing a short circuit.

It is important to remember that lithium-ion batteries are most stressed during charging, and the faster the charging conditions or the cooler the battery temperature, the greater the stress.

The BMS monitors whether the battery is at a chargeable temperature. If the battery is in the chargeable temperature range, charging is allowed. The BMS also controls the charging current when the temperature is low. A well-designed recharging facility first adjusts the battery to room temperature before recharging occurs. This is a good strategy in terms of prolonging battery life.

7.2   Knowing the Condition of a Battery

Lithium-ion batteries are protected inside by a metal case or retort pouch, making it difficult to visually observe the internal state of the battery. I once constructed a battery using a transparent polypropylene bag and observed how the electrolyte was impregnated inside. I called this a skeleton cell. In retort pouches, the laminated aluminium effectively prevents moisture from entering from the outside, but polypropylene alone cannot completely prevent moisture from entering. This is a development technique.

The interesting thing that I observed was the color change of the negative electrode. The graphite of the negative electrode changed from black to red, then to green, and finally to a beautiful golden color as it was recharged. The electrodes face inward against each other, so I think it is more accurate to say that I caught a glimpse of this through a gap. Seeing the sometimes dangerous lithium-ion battery changing to a mysterious golden color inside, which people cannot normally see, I could imagine the lithium-ion battery as a kind of devilish creature. The anode does not emit a golden light, but only changes its state of reflection and absorption of light. If you look closely, you can also observe the anode expanding and contracting.

Putting aside this old story of mine, how can we actually know what is going on inside a battery? To begin with, a battery is an electrical device, and by passing electricity through it, we can learn about its internal state to some extent. For example, it is possible to obtain information about the inside of a battery by charging and discharging the battery at a constant current and following the changes in voltage, or by introducing an alternating current waveform and following the changes in the output waveform. As mentioned in Chapter 1, a battery is a reactor that converts free energy into electrical energy. Since it is known from simple observations of batteries that battery energy is a type of free energy, electrical signals can be used for sensitive sensing of batteries. Other methods have been proposed to precisely capture battery self-heating and entropy changes associated with charging and discharging using an ARC or calorimeter, in terms of heat input/output. These observation techniques are used in practical applications to determine the state of battery deterioration.

7.3 Factors Affecting Battery Degradation

This section repeats some of the information provided in Section 7.1, but the details are important and so are given here.

Before going into the main topic of battery degradation, it is necessary to define why lithium-ion batteries deteriorate and what is meant by the state of battery degradation. Simply put, degradation is a phenomenon in which the amount of charge that can be transferred in and out of a lithium-ion battery decreases, and the resistance of the battery increases, while it is being used. If you have experience in designing and developing lithium-ion batteries, you will have gained the impression that degradation is a very troublesome phenomenon that remains at the end of the process. Various factors are involved in improving degradation resistance (*i.e.* durability), and the properties of the material require the most adjustment in battery design. The next most important factors are the weight ratio of the electrodes and the amount of electrolyte. In addition, there are manufacturing variations, such as foreign matter, water and foreign metal contamination, and initialization defects. During use, high temperatures, overcharging, over-discharging, and charging at low temperatures are the main causes of degradation. What follows is a general discussion of the complete degradation of a lithium-ion battery. Note that other phenomena may be occurring if the design is not well developed.

There are many factors that contribute to degradation, but the first point of degradation of a lithium-ion battery is the surface of the negative electrode. The negative electrode must repeatedly absorb and dissipate lithium ions as it charges and discharges, but the potential at which lithium ions enter and leave graphite is low, at around 0.1 V to 0.2 V relative to the redox potential of lithium, and there are almost no electrolytes or solvents that are durable in such an intense reductive atmosphere (atmosphere of forced electrons). Even those that exist have disadvantages, such as high toxicity and poor ability to dissolve lithium ions. In other words, most solvents are destined to be reduced and decomposed on the anode surface during charging. After reduction decomposition, a film called the SEI forms, which prevents further decomposition of the electrolyte. This process of forming the SEI film is known as initialization.

Although protected by the SEI film, the film is like a scab protecting a wound and does not essentially inhibit electrolyte decomposition on the anode surface. Therefore, even the best lithium-ion batteries will gradually deteriorate as long as graphite is used. The SEI film formed by the decomposition of the electrolyte on the anode grows as the battery degrades. At first glance, one might think that a thicker SEI film would increase the durability of the battery. However, this is not the case. The raw material of the SEI is the electrolyte, so as the SEI grows, it consumes electrolyte in the battery, and lithium ions are trapped as the SEI grows. As explained in the previous section, the basic principle of a lithium-ion battery is to use lithium ions as a charge transport medium. Since the electrolyte is a pathway for lithium ions, the number of transport pathways in a lithium-ion battery will decrease as the amount of electrolyte decreases. Thus, the SEI film protects the surface of the anode, but also acts as an obstacle to the passage of lithium ions. This obstruction to the movement of lithium ions is observed externally as electrical resistance.

Graphite used in lithium-ion batteries has another annoying feature: the expansion and contraction of graphite during charging and discharging. The expansion and contraction of graphite may well induce changes in the shape of the negative electrode and delamination from the collecting electrode. I have already written about why graphite expands and contracts in my explanation of the basic principles of lithium-ion batteries in Section 3.2.1, so I will not go into that here.

Imagine the surface of a rubber ball that is coated nicely, and then air is let in and out of the ball. The surface coating (film) would easily crack. The same thing happens at the graphite surface, although nature is efficient, and any cracks caused by expansion and contraction will be repaired. However, this repair process consumes more electrolyte. In a lithium-ion battery, the electrolyte inside is gradually consumed as the battery is charged and discharged, which in turn reduces the amount of charge that can be transferred in and out of the lithium-ion battery. This degradation process can be summarized as follows.

1. Degradation of the anode surface is the starting point, and lithium ions are trapped in the SEI film.
2. The number of lithium ions that can move freely decreases.
3. The SEI film becomes a resistive body, making it difficult for lithium ions to pass through.

This process explains why the amount of charge (Ah) that a lithium-ion battery can carry in and out over a given time is reduced. This is accompanied by an increase in battery resistance because the resistive film becomes thicker.

This is a phenomenon that occurs in the early stages of degradation. Simply put, the electrolyte is consumed on the surface of the anode, the number of movable lithium ions decreases, and the resistance rises, resulting in a decrease in the charge (Ah) that can be extracted. The phenomenon of the decrease in the number of movable lithium ions can be observed as the development of the SEI and the accompanying decrease in electrolyte. When the electrolyte runs out, rapid degradation of the battery is observed. This is sometimes described as liquid withering.

The process of degradation at the negative electrode eventually extends to the positive electrode. As explained in the basic principles in Chapter 1, the cathode first releases lithium ions when a lithium-ion battery is charged. This release of lithium ions causes the cathode to lose neutrality and "float up", like a submarine with blown ballast (the direction of sinking is based on electron volts, but it is complicated, so I will not discuss that here). This phenomenon is observed as an increase in voltage accompanying charging (actually, the voltage is the difference in potential, so the expression is incorrect, but the movement of the cathode side matches the movement of the battery, so I will use that description). During discharge, lithium ions are sucked back into the cathode so the cathode absorbs lithium ions and "sinks" again. This is observed as a drop in voltage (since the change in potential of the cathode is in the same direction as the change in voltage of the battery). If the number of lithium ions that can move is reduced, the number of lithium ions that can return to the cathode during discharge is also reduced, meaning that the cathode cannot "float" to its original location. The cathode has areas where capacitance can be obtained and areas where capacitance cannot be obtained, depending on the potential. Designers naturally aim to maximize capacity by balancing the positive and negative electrodes. However, if the number of lithium ions that can return to the cathode during discharge is halfway, the cathode will fall outside the area where maximum capacity can be generated. This is called the "capacity gap" of the cathode. This is a strange expression, but it is actually a phenomenon in which the cathode deviates from the (maximum) capacity generating area that was initially set. It is also known that if lithium ions that are supposed to return to the cathode do not return, the crystals themselves are damaged. Thus, in addition to the three degradation factors that originate from the negative electrode, the cathode's capacity shift, whereby the capacity of the cathode cannot be used properly, accelerates the degradation of the lithium-ion battery. Eventually, the battery will reach the end of its life due to liquid depletion.

So, is it possible to extend battery life by increasing the amount of electrolyte? The answer to this question is "yes, to some extent". It is possible to extend the time until rapid degradation by increasing the amount of electrolyte. However, this depends on the cost, and an increase in the amount of electrolyte would be expected to increase the combustion power in the event of an accident.

As shown in Figure 7.1, as a battery deteriorates, its resistance increases and its capacity decreases. On the other hand, the voltage

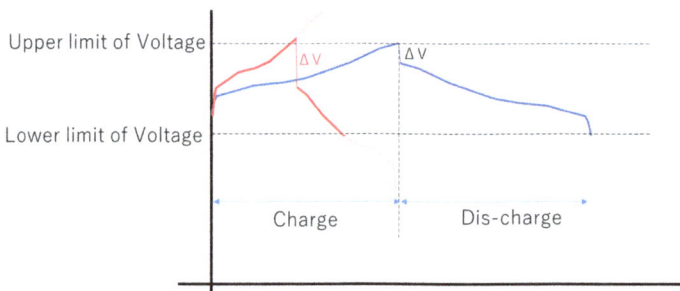

Figure 7.1 Schematic diagram comparing battery charge/discharge curves before and after degradation.

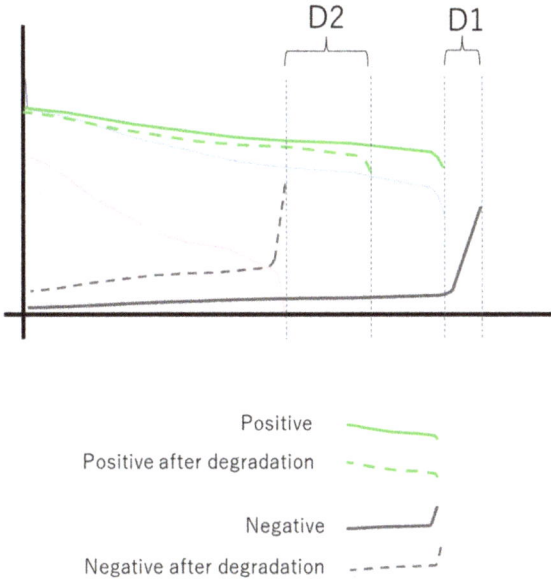

Figure 7.2   Change in waveform of the positive and negative electrodes before and after cell degradation.

range over which the battery operates remains the same, so the range over which the battery capacity can be used decreases. Since the charge–voltage curve changes in a way that reflects the deterioration and capacity decreases, the state of battery deterioration can be estimated from the shape of the curve.

Figure 7.2 shows the potential change between the positive and negative electrodes inside a battery. Normally, the capacitance scale of the positive and negative electrodes is designed with that of the negative electrode being slightly larger (D1). As degradation progresses, the utilization ratio of the positive and negative electrodes changes (D2) due to their different degrees of degradation. The difference between D1 and D2 is referred to as capacitance misalignment.

7.4   Prediction of Degradation Is Based on a Map

The most commonly used method for understanding the degradation trend of batteries is to use a graph with the number of cycles, operating hours, or total Ah on the horizontal axis, and the battery's capacity retention rate or change in resistance on the vertical axis. Rather than using complicated measurements and calculations, it is simpler and more reliable to actually degrade the battery and observe the trends.

As a starting point, we should define the temperature range in which lithium-ion batteries work, which differs between the charging side and the discharging side. The discharging side can ideally be used in the temperature range of 10 °C to 45 °C, or even −10 °C to 60 °C with a little effort. The fact that lithium-ion batteries generate heat when discharged must be taken into account beforehand. On the charging side, a temperature range of 25 °C to 45 °C, or even 10 °C to 60 °C if slightly overworked, is appropriate. The starting point of degradation is the expansion and contraction of lithium ions entering the negative electrode during charging, and since this tendency is accelerated by charging at low temperatures, the temperature conditions on the charging side must be strict. When charged at a current greater than that specified, the potential partially decreases below the lithium deposition potential on the anode surface, which results in precipitation of metallic lithium on the anode surface and can induce an internal short circuit or other hazardous conditions. This tendency also accelerates at low temperatures. In general, manufacturers recommend the optimal discharge or charge rate depending on the temperature.

Let us consider mapping the degradation state of a battery under normal use. For the time being, it would be good to collect cycle degradation characteristics at 10 °C, 25 °C, 40 °C, and 60 °C. The so-called "storage degradation", when a battery is stored with a full charge, is different from the "cyclic degradation" in terms of the degradation mode. The reason for this is that the storage degradation mode does not involve expansion and contraction of the negative electrode. It is advisable to make observations in both modes. In general, Arrhenius's law (an empirical law that states that degradation doubles every 10 °C) is followed in the storage degradation mode, while in the cyclic mode the degradation is often proportional to time or the number of cycles (or "route"). In some cases, the "route" is a straight line, and in other cases it is not necessary to take that route. As a rule of thumb, if degradation depends on both poles, it is better to take the cycling "route".

The following discussion is limited to cyclic degradation. The graph shown in Figure 7.3 is obtained by taking the number of cycles (the route of the cycle) on the horizontal axis and the capacity retention rate on the vertical axis. The degree of degradation tends to increase as the environmental temperature increases. Using Excel, you can easily add an approximation curve, and the square value of the relative error, indicated by R^2, is the distance from the approximate straight line to the actual data. The closer this is to 1, the smaller the approximation error, and an R^2 value of 0.97 or more indicates high reliability

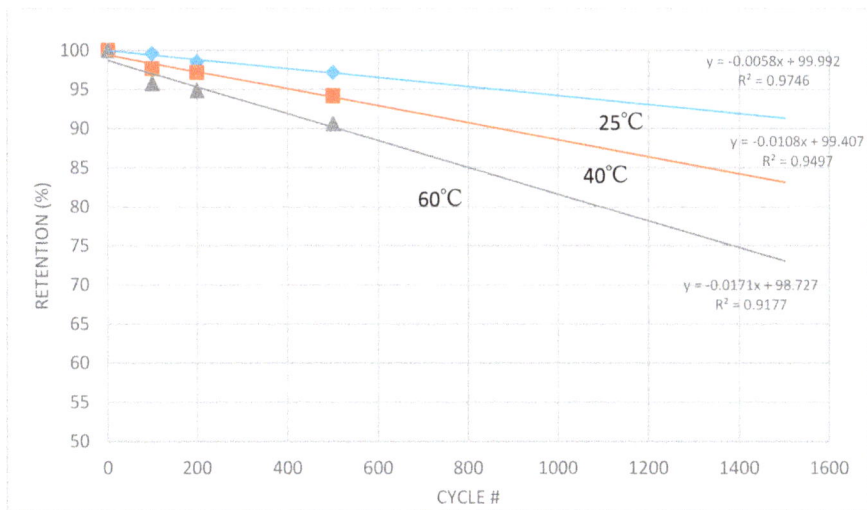

Figure 7.3 Recorded data for the actual degradation of LFP-graphite batteries.

for the data. The use of the square (in R^2) comes from the fact that the distance from the approximate curve to the actual data (*i.e.* the relative error) is expressed using the Pythagorean definition (*i.e.* area).

If the slope at 60 °C is within a factor of 2 of that at 40 °C, the battery is considered to have degraded within expectations over its operating temperature range (here, 25–60 °C) as indicated by the first-order reaction formula. The battery is considered to be degraded at temperatures other than those measured, *e.g.* 45 °C, with a similar trend. This means that, based on the approximate straight lines for 25 °C, 40 °C, and 60 °C, we can describe the degradation of the battery over the whole temperature range (*i.e.* 25–60 °C). This data can be used as a map.

Normally, the Arrhenius plot shows the time taken to fall below the reference capacity *versus* capacity, but, in this case, the horizontal axis is the number of cycles, for simplicity. Therefore, when calculating the energy of degradation, it is necessary to convert the plot into a time-dimensional one.

The approximate curve can be conveniently extrapolated forward. A note, however, on how much to trust this. As mentioned earlier, as lithium-ion batteries deteriorate, they eventually reach a state called "liquid exhaustion", and their capacity deteriorates rapidly. Extrapolation can predict, to a certain extent, the state of degradation before the liquid runs dry. However, once liquid exhaustion occurs, a different mode of degradation occurs, which cannot be predicted well based on data obtained up to that point.

The method used by battery users to determine the current level of deterioration is to count the amount of charge (Ah) stored during charging and discharging, and to compare this with the previous charge capacity. This trend is compared using a map to determine the current degree of deterioration. Of course, since the amount of charge changes according to the previous discharge amount, the calculation should be based on the balance of charge between discharge and charge. This is called Coulomb counting. Since how the battery is used is different for each user, there is a possibility of deviation from Coulomb counting alone, and this is corrected using the map.

New batteries are often designed one after another at the development site, and there may not be time to create a map from actual data. In such cases, for example, accelerated tests may be conducted at the maximum operating temperature of the battery in question, say 60 °C, and the results may be reflected in the temperature below the maximum temperature.

The SEI film on the anode surface has a significant effect on the resistance of the battery. The resistance of the SEI film tends to increase with temperature, especially at low temperatures. For this reason, it is necessary to periodically remove the battery from the constant-temperature bath and measure it at a predetermined temperature (usually 25 °C) rather than measuring it while it is still in the constant-temperature layer.

7.5   Understanding Internal Conditions Using Impedance

The term "impedance" may seem somewhat confusing to those who work in the field of pure chemical synthesis. This used to be the case for me. However, once you learn about battery degradation estimation technology based on impedance, you will realize that it is a very useful technology, especially for those who develop and design batteries.

The term impedance is defined as the ratio of voltage to current (voltage/current) in a phasor representation in an AC circuit. Phasor representation simply means that it is expressed using complex numbers. You might feel tempted to look away at the mention of complex numbers, but don't worry. In this case, complex numbers are just being used to forcefully "plot" points that can't be represented on the same graph. Nowadays, computer software handles the calculations. Just think of it casually as, "Oh, so complex numbers are being used".

Putting this representation aside, since impedance is the ratio of voltage and current in an AC circuit I would like to simply call it "resistance", but that is not the case. The definition of resistance is, of course, determined by Ohm's law: $\delta E = \delta IR$, which is expressed as the ratio of voltage/current over time. However, in the case of alternating current, some elements work similarly to resistance, but are not resistance as it is defined by Ohm's law. These are capacitance and reactance.

The AC waveform is a regular wave, like a sine wave. Suppose that when an alternating current is passed through a certain device, nothing changes in the waveform. If only resistance were present, the wave would be smaller overall. But, what if the shape of the wave is distorted? One possible cause of distortion of the wave shape is impedance, which includes capacitance and reactance. Inside a battery, countless paths (circuits) exist for electricity to flow. Imagine two circuits. One circuit has a component that temporarily stores electricity in the middle, while the other does not. In the first circuit, electricity is stored in this component before being released. This causes a delay. The other circuit lacks such a component, so no delay occurs. Since electricity is a wave, the delayed wave and the undelayed wave will interfere with each other, likely causing distortion or delay in the waveform. The electrical circuits that make up a battery primarily contain two types of components that store electricity: capacitance and reactance. The former is electrical energy stored in capacitors, and the latter is electrical energy stored in coils. Here, when the wave slows down compared to the original waveform, it is called "slowing down", whereas the phenomenon where the wave advances is called "advancing". Impedance is simply a generic term for resistance-like components in an alternating current. The observed result is a composite of resistance, capacitance, and reactance components, all of which can distort the input waveform.

Since alternating current refers to the regular transmission of electricity, it can be simply stated that any electricity in which the voltage varies with time is alternating current. When a switch is placed in the middle of a circuit and opened and closed periodically by a timer, a square wave is obtained, which is, of course, a type of alternating current. Because it is a type of alternating current, a square wave also has impedance. What happens then if we add reactance? The moment

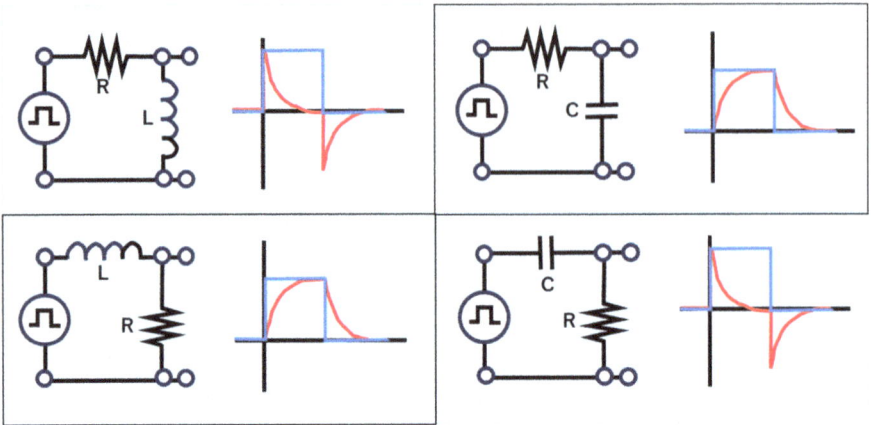

Figure 7.4   The combination of resistive and capacitive or inductive components distorts the original input waveform, as shown.

the switch is turned off, an upswing in the waveform will be observed
(Figure 7.4).

Batteries also have capacitance and reactance, which affect the
charge–discharge waveform of the battery. By applying alternating
current to a battery and observing its response (waveform change),
changes in the resistance and capacitance components inside the battery can be observed.[11]

By looking at changes in battery impedance, you can tell how much
the battery has deteriorated (Figure 7.5).

If the battery had no impedance component, it would output a voltage waveform exactly analogous to the input current. However, the
output waveform would be distorted, as shown in Figure 7.5, due to
various delays and advances caused by the impedance. By analyzing
this distortion in various ways, we can estimate the components that
make up the impedance circuit in the battery.[12]

When alternating current with different amplitudes is continuously
applied to a battery, plotting the input frequency on the horizontal axis
and the actual observed frequency on the vertical axis produces a plot
called a Cole–Cole plot (Figure 7.6). The Cole–Cole plot can be read to
show the effects of the resistance, capacitance, and reactance components built into the battery. Figure 7.6 shows the results of actual measurements on an 18650-type cylindrical cell. Since impedance varies
greatly with the SOC, the SOC to be measured must be determined
from the beginning; the higher the SOC, the lower the impedance

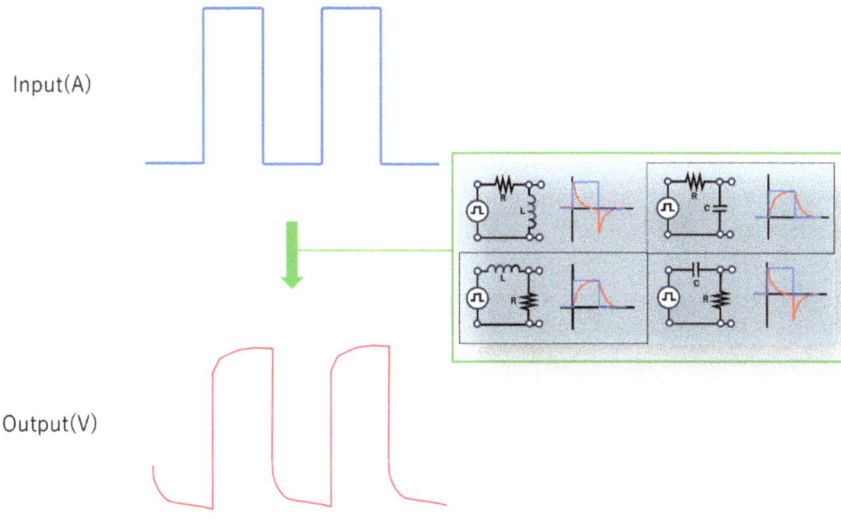

Figure 7.5 Inside the battery, there are numerous capacitance and reactance components. When AC current flows through it, various waveform distortions and delays are observed. Analyzing these waveforms reveals the battery's state of degradation.

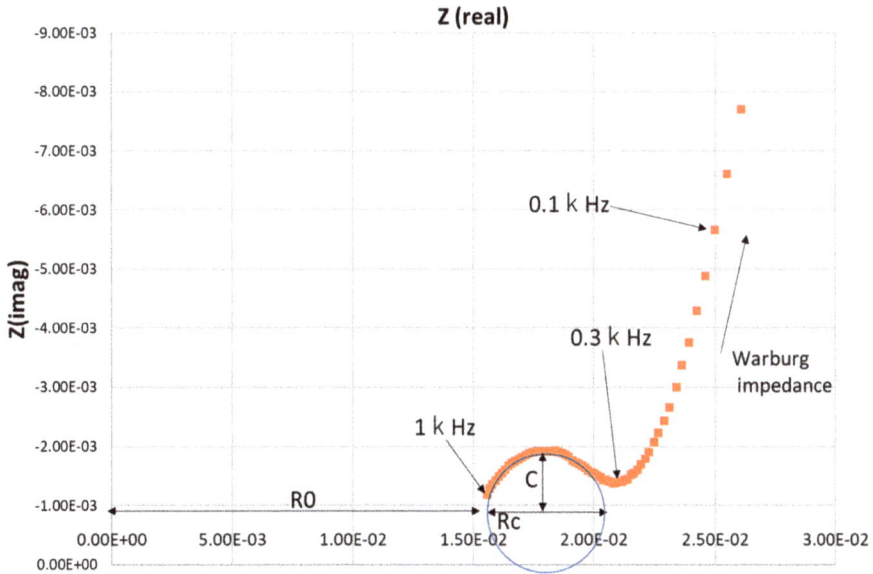

Figure 7.6 Example of a Cole–Cole plot for a cylindrical lithium-ion battery.

tends to be. This is mainly because the negative electrode has less resistance to absorb lithium ions. It is said that this is because the carbon becomes more metallic (because it contains more lithium).

Simply put, the distance from zero to the intersection of the arc and the *x*-axis corresponds to the resistance, the height of the arc represents the capacitance component, and the width of the arc represents the resistance component that is parasitic on the capacitance.

In the past, researchers used to print out a Cole–Cole plot and separate the above components by applying a ruler to it. However, this was tedious and time-consuming and not practical. More recently, good fitting software has become available.

In order to analyze a Cole–Cole plot, it is necessary to create a virtual circuit model of the circuits that are contained in the battery to perform the fitting process.[13] When I have done this previously and manually, I assumed a Randolph-type circuit with only one circuit. Recently, as a rule of thumb, it seems to be possible to obtain results close to the actual situation by fitting a virtual circuit called a four-ladder Randolph-type (Figure 7.7), which consists of about four Randolph-type circuits connected in series, inside the battery. Of course, the more circuits there are, the more complicated the fitting regression calculation becomes, so a computer program is a prerequisite. If you install a testing device capable of frequency analysis, the software is usually included.

In the structure of a battery, it is the collecting electrode that generates reactance. As you can imagine, it is unlikely that the collecting electrode immediately contributes to battery degradation. In addition, the reactance component decreases the information on the high-frequency side. In other words, when measuring batteries with large reactance, the information on the high-frequency side becomes unreliable. In particular, cylindrical batteries should be noted because the contribution of reactance is large due to their structure. The reactance component is also a parasitic component of the cables that

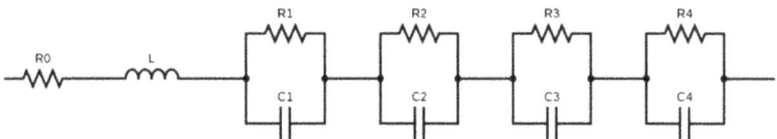

Figure 7.7 Example circuit for calculation results when fitting a hypothetical four-ladder equivalent equation.

Table 7.1 The component constants analyzed by using the equivalent circuit model of the four-ladder type in the software that comes with the impedance analyzer.

Test 1	R0 + L0 + R1//C1 + R2//C2 + R3//+R3//C3 + R4//C4
R0	0.0253412
L0	2.98×10^{-38}
R1	0.04687
C1	0.30191207
R2	0.00464445
C2	46.1467831
R3	0.02296107
C3	316 236 305
R4	0.00315481
C4	2.637986

make up the measurement circuit, so it is necessary to use cables recommended by the manufacturer when recording a good Cole–Cole plot.[14]

Let us calculate the circuit constants for the circuit in Figure 7.7, with the results as shown in Table 7.1.

It is said that R0 (Rs in other circuits) is the battery's solid-state diffusion resistance, which is equal to the DC internal resistance, Ladder 1 is the CR (Capacitance–Resistance) time constant[†] in the presence of SEI, Ladder 2 is the positive pole, Ladder 3 is the negative pole, and Ladder 4 is the CR contribution due to diffusion. For example, if the trends of R0 and R4 match, it can be assumed that R4, or diffusion, contributes to the overall battery resistance increase. In addition, since C1 is said to indicate SEI capacity, the trend in SEI film increase can be estimated from the increase in C1. As explained earlier, battery degradation begins with an increase in the anode SEI film, so if the movement of R1 and C1 is known, the data are significant.

The Cole–Cole plot sometimes reveals that the waveforms separate into two as the battery deteriorates, as shown in Figure 7.8. The waveform is hat-like. This indicates that the reaction of the positive and negative electrodes to AC is becoming separated. The answer to the question "Which electrode is it?" depends on whether the cause of the degradation is actually in the cathode or the anode. In general, one can

---

[†]When alternating current flows through a circuit containing a capacitor and a resistor, a delay occurs in the output waveform depending on the values of the capacitance and resistance. The combination of capacitor and resistor that causes this delay is called the CR time constant. By understanding the CR time constant, it is possible to estimate the values of the capacitance and resistance through reverse calculation.

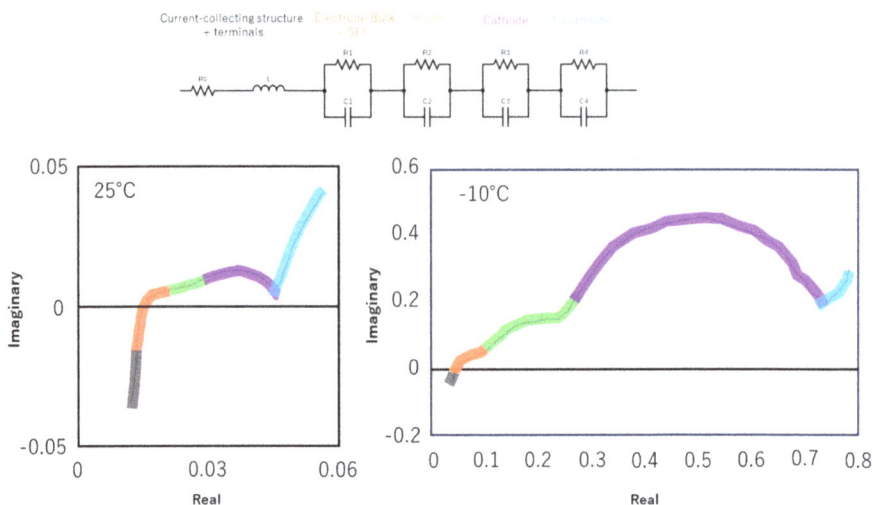

Figure 7.8   Fitting using four ladders. To highlight the built-in resistance, measurements are taken at −20 °C, a temperature at which the internal resistance of the battery increases. As the battery deteriorates and the resistance increases, the contributions of the positive and negative electrodes can be clearly separated, as shown.

envision a case in which the anode resistance is initially so small that it is hidden by the cathode resistance, but as the SEI film increases and the anode resistance increases, they separate and become distinguishable. By comparing the circuit constants measured in advance according to the battery's degradation level with the actual measured circuit constants, the battery's degradation state can be estimated.

7.6   Analyzing Battery State Using Charging and Discharging Waveforms (dV/dQ Analysis)

Batteries must be recharged before they can be used. Since there is always a time to recharge a battery in a BEV, even when considering its use in a car, a method to determine the state of battery deterioration based on changes in the charging waveform is convenient for determining the state of battery deterioration during use. As explained, as a lithium-ion battery is used, the SEI on the anode surface becomes thicker, and the resistance of the battery gradually increases because the SEI is a component of resistance to the movement of lithium ions. Also, as the total amount of lithium ions that can be transferred decreases, the amount of charge that can be transferred in unit time decreases. Since the solvent that makes up the electrolyte is also consumed, the number of transport routes through which lithium ions can pass decreases. Together, these factors increase the resistance of a lithium-ion battery, and since lithium ions are trapped in the SEI, not enough lithium ions can return to the positive and negative electrodes during charging and discharging, and the capacity of the battery decreases. The battery charge waveform will then change as the battery deteriorates, as shown in Figure 7.8. Generally, charging is performed within a limited temperature range specified by the battery manufacturer. The charge curve can be collected and stored under typical temperature conditions covering this temperature range in advance, and then used as a map.

The dV/dQ is another technique for estimating battery degradation based on charge (discharge) waveforms. However, as the equation form suggests, it uses the amount of voltage change per unit capacity, *i.e.* the differential waveform. Figure 7.9 shows a typical example. In the case of lithium-ion batteries using graphite, changes in differential values accompanying clear changes in the stage structure, especially at the negative electrode, can be obtained.

A little explanation is in order here. Previously, it was explained that the graphite anode intercalates lithium ions at a specific potential and transitions to the state represented by C_6Li (see Chapter 3). In fact, it is known to take the form of $C_{12}Li$, $C_{18}Li$, *etc.*, before reaching C_6Li.[15] For such stage changes, there is a clear potential change, as shown in Figure 7.9.

In the stage leading to C_6Li, the potential is low and the reduction force is strong, so SEI formation on the negative electrode in question proceeds, and, where the SEI film becomes thicker, it becomes difficult for lithium ions to move and be used. Therefore, the C_6Li stage appears to deteriorate and lose capacity as a phenomenon visible in the table. The good thing about dV/dQ analysis is that the cathode also has a specific stage change, so the degradation state of the cathode or anode can be identified separately.[14]

7.7 Relationship Between Degradation and Safety Inferences Based on ARC Data

There seems to be a delay in discussing the safety of degraded batteries. When I was involved in the design and development of batteries for use in vehicles, it was my clear understanding that the fresher a

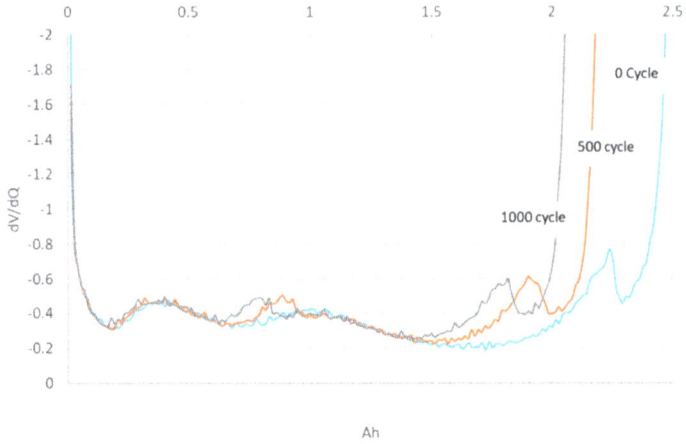

Figure 7.9   Change in dV/dQ curve with degradation progression. The degradation state is ascertained based on the ratio of the change in the peak associated with the phase change of the anode to the change in the peak associated with the phase change of the cathode. It can be easily calculated based on the waveform of a battery during charging.

lithium-ion battery is, the greater the risk of a hazardous situation. This is simply because the battery contains a large amount of electrolyte, which becomes fuel in the event of an accident, and it was thought that the combustion power of a near-dead cell would be weaker. In fact, the trend seems to be that fresh cells generate more heat at the time of an accident. ARC was used to verify this point by accurately capturing the self-heating that leads to thermal runaway of lithium-ion batteries (Figure 7.10).

The results complement the above rule of thumb. However, self-heating starts at lower temperatures in degraded batteries, indicating that the overall thermal stability of the batteries is lower. This tendency is more pronounced for batteries that have been cycled at low temperatures, with some batteries starting to self-heat at around room temperature.

Generally, the older a lithium-ion battery is, the lower the damage tends to be in the event of an accident, but it should be noted that the older the battery is, the more it begins to self-heat at lower temperatures.

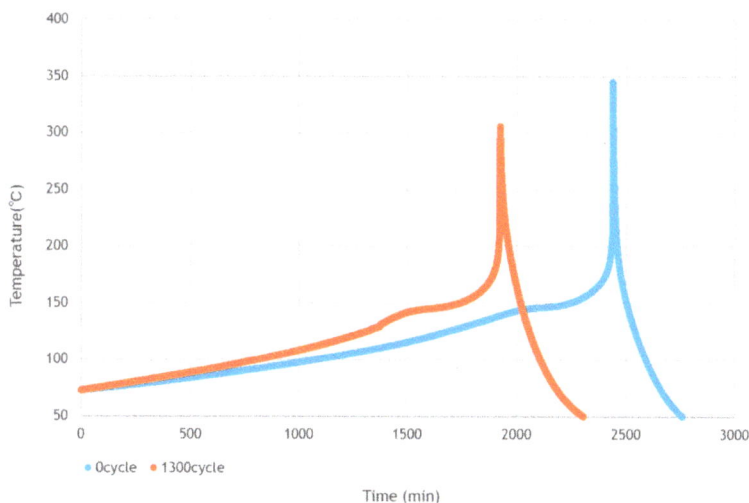

Figure 7.10 Variation of the thermal runaway ARC curve of a typical lithium-ion battery cell before and after degradation. The time to thermal runaway becomes shorter as degradation progresses. At the same time, the maximum peak of heat generation decreases.

7.8 Closing Remarks

This chapter has covered the principles of lithium-ion battery degradation and the analysis of degraded states at a fairly rapid pace. While the content may have been difficult for beginners entering the field of lithium-ion batteries to grasp, we have illustrated representative analysis methods. Some readers may feel they haven't fully digested the material. Please refer to the papers cited and continue your studies.

References

1. M. Hirooka, T. Sekiya, Y. Omomo, M. Yamada, H. Katayama, T. Okumura, Y. Yamada and K. Ariyoshi, Degradation mechanism of $LiCoO_2$ under float charge conditions and high temperatures, *Electrochim. Acta*, 2019, **320**, 134596.
2. M. Nishijima, T. Ootani and Y. Kamimura, *et al.*, Accelerated discovery of cathode materials with prolonged cycle life for lithium-ion battery, *Nat. Commun.*, 2014, **5**, 4553.
3. T. Hayashi, J. Okada, E. Toda, R. Kuzuo, N. Oshimura, N. Kuwata and J. Kawamura, Degradation mechanism of $LiNi_{0.82}Co_{0.15}Al_{0.03}O_2$ positive electrodes of a lithium-ion battery by a long-term cycling test, *J. Electrochem. Soc.*, 2014, **161**(6), A1007.
4. H. Shin, Y. K. Lee and W. Lu, Structural degradation of graphite anode induced by dissolved manganese ions in lithium-ion batteries, *J. Power Sources*, 2022, **528**, 231223.

5. C. Birkl, M. Roberts, E. McTurk, P. Bruce and D. Howey, Degradation diagnostics for lithium ion cells, *J. Power Sources*, 2017, **341**, 373–386; M. Shikano, H. Kobayashi, S. Koike, H. Sakaebe, Y. Saito, H. Hori, H. Kageyama and K. Tatsumi, X-ray absorption near-edge structure study on positive electrodes of degraded lithium-ion battery, *J. Power Sources*, 2011, **196**(16), 6881–6883.

6. Y. Kobayashi, H. Miyashiro, A. Yamazaki and Y. Mita, Unexpected capacity fade and recovery mechanism of $LiFePO_4$/graphite cells for grid operation, *J. Power Sources*, 2020, **449**, 227502.

7. J. S. Edge, S. O'Kane, R. Prosser, N. D. Kirkaldy, A. N. Patel, A. Hales, A. Ghosh, W. Ai, J. Chen, J. Yang and S. Li, Lithium ion battery degradation: what you need to know, *Phys. Chem. Chem. Phys.*, 2021, **23**(14), 8200–8221.

8. S. E. O'Kane, W. Ai, G. Madabattula, D. Alonso-Alvarez, R. Timms, V. Sulzer, J. S. Edge, B. Wu, G. J. Offer and M. Marinescu, Lithium-ion battery degradation: how to model it, *Phys. Chem. Chem. Phys.*, 2022, **24**(13), 7909–7922.

9. T. Yoshida, M. Takahashi, S. Morikawa, C. Ihara, H. Katsukawa, T. Shiratsuchi and J. I. Yamaki, Degradation mechanism and life prediction of lithium-ion batteries, *J. Electrochem. Soc.*, 2006, **153**(3), A576.

10. J. Zhu, M. S. Darma, M. Knapp, D. R. Sørensen, M. Heere, Q. Fang, X. Wang, H. Dai, L. Mereacre, A. Senyshyn and X. Wei, Investigation of lithium-ion battery degradation mechanisms by combining differential voltage analysis and alternating current impedance, *J. Power Sources*, 2020, **448**, 227575.

11. https://www.global.toshiba/content/dam/toshiba/migration/corp/techReviewAssets/tech/review/2018/06/73_06pdf/f01.pdf.

12. K. Das and R. Kumar, Electric vehicle battery capacity degradation and health estimation using machine-learning techniques: a review, *Clean Energy*, 2023, 7(6), 1268–1281; S. Okada, S. Yoshitake, Y. Tominaga and A. Anekawa, Development of Lithium-ion Battery Deterioration Diagnosis Technology, *Yokogawa Tech. Rep. Engl. Ed.*, 2013, **56**(2), https://web-material3.yokogawa.com/1/11298/tabs/rd-te-r05602-007.pdf.

13. C. Zhu, L. Sun, C. Chen, J. Tian, W. Shen and R. Xiong, Lithium-ion battery degradation diagnosis and state-of-health estimation with half cell electrode potential, *Electrochim. Acta*, 2023, **459**, 142588.

14. S. Taminato, M. Yonemura and S. Shiotani, *et al.*, Real-time observations of lithium battery reactions—operando neutron diffraction analysis during practical operation, *Sci. Rep.*, 2016, **6**, 28843.

15. I. Umegaki, *et al.*, *Phys. Chem. Chem. Phys.*, 2017, **19**, 19058–19066.

8 Future Prospects for Lithium-ion Batteries

In this final chapter, I consider how much further the performance of lithium-ion batteries can be improved and whether ion species other than lithium ions can be used as "ion" batteries. The market outlook for lithium-ion batteries from a technological perspective is also discussed.

Advances in the world of batteries are extremely rapid. Recently, batteries based on concepts like anode-less designs have been developed. However, this book has focused on explaining so-called traditional lithium-ion batteries. Therefore, this chapter will explain technologies surrounding traditional lithium-ion batteries or those slightly beyond them. In a lithium-ion battery, electrons move between the positive and negative electrodes during charging and discharging, and lithium-ions also move behind the electrons to compensate for the charge. The positive electrode mainly uses a metal oxide that contains lithium-ions, and the negative electrode uses carbon. As explained in Chapters 2 and 3, this is the basic operation and structure of a lithium-ion battery. With the metal oxide system currently in use, the theoretical capacity is fixed at around 270 mA h g^{-1}, and, considering the balance between lifespan and safety, around 200 mA h g^{-1} can be used. The ratio of actual capacity to theoretical capacity is called the utilization rate, and this was explained in Chapter 3. Currently, a realistic approach to improving performance on the positive electrode

RSC Foundations No. 4
Li-ion Batteries
By Hitoshi Nakamura
© Hitoshi Nakamura 2025
Published by the Royal Society of Chemistry, www.rsc.org

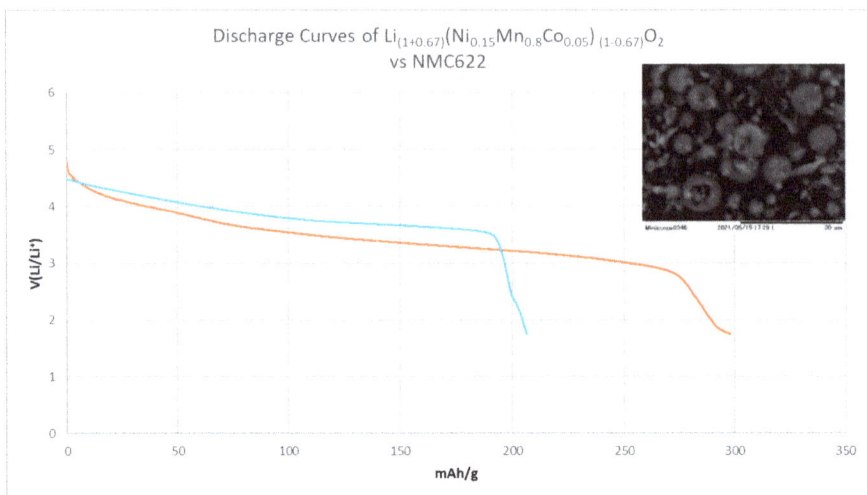

Figure 8.1 Discharge curve of $Li_2Mn_2O_3$–$LiNiO_2$ hybrid cathode material, reaching a maximum capacity of 300 mA h g^{-1}, which is larger than for NMC622.[2]

side is to further increase this utilization rate. There are already some limits to the ingenuity that can be used to make better use of materials. As explained in Chapter 4, this is because there is a concern that increasing the utilization rate will compromise reliability and safety. It seems that we will soon need to look for a different material system. There are materials that can break through the 270 mA h g^{-1} barrier. One of these is the Li_2MnO_3 system. This material contains two lithium atoms, so the theoretical capacity is calculated by dividing twice the theoretical capacity of 26.8 mA h (as explained in Chapter 3) by the molecular weight of 117. In other words, $26.8 \times 2/117 = 451$ mA h g^{-1}. This is a very high theoretical capacity compared to the layered ternary systems currently in use. However, Li_2MnO_3 itself is insulating, so it is used in a hybrid form with conventional $LiMO_2$ (where M is a metal).[1] In metallurgy, the term "hybrid" is not used, so the technical term is "solid solution". This means that both components are dissolved together. For example, when Li_2MnO_3 and $LiNiO_2$ are combined in a ratio of $8:2$, a positive electrode material with a high theoretical capacity of around 340 mA h g^{-1} is obtained (Figure 8.1). The synthesized hybrid material contains 1.67 lithium atoms within each molecule, and during actual discharge, it shows a capacity of 300 mA h g^{-1}, which is higher than that of NMC622, a type of layered ternary oxide material.

On the other hand, to increase the capacity of the anode, a method of mixing silicon with graphite is used. Silicon has a capacity of around 4200 mA h g^{-1} and has a low potential for storing lithium ions of around 0.5 V. However, when silicon stores lithium ions, it expands, which raises issues of durability and safety. For this reason, various measures have been used to reduce this expansion, such as combining silicon in the form of nanoparticles with conventional graphite, and the capacity of the anode has been gradually improved.[3]

By combining the positive and negative electrode materials introduced here, it is expected that lithium-ion batteries with a maximum energy density of around twice that of current commercial products can be achieved.

As mentioned in Chapter 2, and based on the research that led to the 2019 Nobel Prize, a similar battery that uses lithium ions as a charge transfer medium but does not use carbon in the anode does not fall under the definition of a lithium-ion battery. For this reason, I have not mentioned this previously in this book. Here, however, I will leave aside the circumstances of the invention of lithium-ion batteries and expand the definition to simply "a battery that uses lithium-ions as a charge compensation medium".

The candidate for a cathode material with the largest capacity found so far is sulfur. Sulfur has a huge capacity of around 1675 mA h g^{-1}. It is also cheap. The ultimate anode, however, is metallic lithium, which has a huge capacity of 3860 mA h g^{-1}. If these two are combined, it is possible to create a battery with a capacity of around 1500 mA h g^{-1} that generates a voltage of around 2 V. If a stable battery can be created using this combination, it will be possible to create a lithium–sulfur (LiS) battery with an energy density around 3–4 times that of current lithium-ion batteries. There are issues with LiS batteries, however. One is that sulfur dissolves into the electrolyte during redox reactions. There are two ways to overcome this, one of which is to use certain types of ether or molten salt as the electrolyte. In 2011, a team led by Dr Masayoshi Watanabe of Yokohama National University in Japan demonstrated that stable operation was possible by using glyme, which became a major topic of conversation.[4] The other method is to use a solid electrolyte, and various attempts have been made.[5] The method of using metallic lithium as the anode was attempted before the lithium-ion battery was introduced to the world, as explained in Chapter 2. However, at the time, there were issues with durability and safety, so lithium-ion batteries using carbon instead of metallic lithium were developed. However, the huge capacity of metallic lithium is attractive, so efforts to make use of metallic lithium continue.[6] Changing the electrolyte is a key point in effective methods for making use of metallic lithium, and the application of molten salts or solid electrolytes is also effective. Rechargeable lithium batteries that use metal anodes have low durability and reliability, and in the past they were replaced by lithium-ion batteries due to the high incidence of fires, so it is important to reconfirm that they are not a simple technology to master.[7]

I believe the above modifications are important in terms of the outlook for performance. However, for release to the world, reliability and safety in particular need to be guaranteed. In recent research, batteries that do not require a negative electrode have been attracting attention. They are a type of metal lithium battery, but in reality, the negative electrode is only generated during charging. When discharging, this disappears cleanly, and the battery becomes negative electrode-free. When observed during discharge, the negative electrode has gone.[14] The electrolyte is important for the realization of such a battery, and a type that uses a solid electrolyte has been proposed. I think that this is an important technology for the preliminary stage of the all-solid-state or molten-salt LiS battery.

As I am not an economist, I will provide my outlook on the future market from a technical perspective.

As you know, the Earth's climate appears to be losing its balance. One of the causes of this is the involvement of greenhouse gases, particularly CO_2.[8] For this reason, there is a worldwide movement to reduce the use of fossil fuels and promote the use of renewable energy. One of the most prominent recent developments is the shift from fossil fuels to electric power as the propulsion energy for automobiles. When comparing the systems of internal combustion engines and electric vehicles, electric vehicles are simply more energy-efficient. However, the final conversion efficiency is affected by the source of the electricity, and electric vehicles achieve their highest energy conversion efficiency when using renewable energy.[16] Renewable energy from natural sources, such as solar and wind power, aims to make effective use of energy that was previously unused, such as sunlight and wind power. Rather than the stable power output that can be obtained by burning fossil fuels, natural renewable energy is dependent on the movements of nature, and output fluctuates. For this reason, a system that stores energy, such as in a battery, is needed and allows stable and effective use. It may sound a little extreme, but global warming is a pressure being applied to humanity by the Earth itself, so it is obvious that the industries that can guarantee to reduce this pressure, namely the electric vehicle and renewable energy industries, will continue to grow in size. In particular, it is predicted that expectations for the battery industry, which will become a common key device for renewable energy in the future, will continue to grow.

I would also like to mention some developments other than lithium batteries. As explained in Chapters 1 and 2, lithium-ion batteries use lithium ions as a charge transfer medium to compensate for the movement of electrons. Historically, hydrogen was used as an ion to compensate for the movement of electrons, in Volta's battery and Planté's lead storage battery, before lithium, which is slightly heavier than hydrogen, was used. Research is also underway to see whether sodium and potassium, which are below lithium in the periodic table, can be used. The development of sodium-ion batteries is currently undergoing a second boom, following the announcement of mass production by a major Chinese battery manufacturer in recent years. The operating principle of sodium-ion batteries is basically the same as that of lithium-ion batteries, with the only difference being that lithium ions are replaced with sodium ions. For this reason, the same manufacturing process as lithium-ion batteries can be used. In addition, the fact that sodium is more abundant on Earth than lithium and is cheaper and easier to obtain is also a factor.[9] Lithium is unevenly distributed in terms of where it is found, but sodium can be extracted anywhere on Earth. According to my research, investigations into sodium-ion batteries began before the invention of lithium-ion batteries.[10] In 2000, it was reported that hard carbon could be used in the negative electrode.[15] Later, Sumitomo Chemical in Japan submitted a patent[11] for a full cell that uses metal oxide in the positive electrode and hard carbon in the negative electrode, and reported on the operation of the cell,[12] and Dr Komaba of Tokyo University of Science succeeded in operating a full cell of a sodium-ion battery in 2011.[13] At present, it appears that the energy density is inferior to that of lithium-ion batteries. However, according to the latest research results, the energy density of sodium-ion batteries appears to have reached the same level as that of lithium-ion iron phosphate batteries (Figure 8.2), which are widely used in BEVs.

There also reports that sodium-ion batteries, while not completely safe, are much safer than lithium-ion batteries.[17] As explained earlier, the range of applications for "ion" batteries in equipment is expected to increase due to pressure from the Earth, so I think there is a good chance that batteries that use sodium ions will also become established.

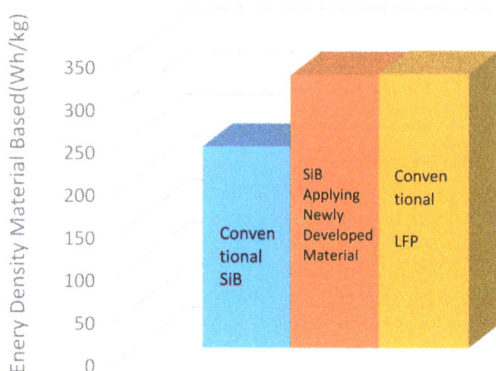

Figure 8.2   Chart comparing the weight energy density of LFP and sodium-ion batteries (SiB) based on their materials. It shows that it is possible to construct a sodium-ion battery with an energy density equivalent to an LFP battery using recently developed cathode active materials.[18] Reproduced from ref. 18 with permission from the Author.

Protecting the global environment and improving battery performance are closely linked. It is a good thing that we are gradually moving away from using fossil fuels. As this trend continues, the scale of production of lithium-ion batteries will continue to increase, and various research and development approaches will be needed to improve their performance. As a scientist, I am very interested in preserving the global environment. I think it would be wonderful if the research you may be about to start could help to lower the temperature of the Earth, even a little, and I would be very happy if this book helps you to learn about lithium-ion batteries.

References

1. M. Tabuchi, Y. Nabeshima, T. Takeuchi, K. Tatsumi, J. Imaizumi and Y. Nitta, Fe content effects on electrochemical properties of Fe-substituted Li_2MnO_3 positive electrode material, *J. Power Sources*, 2010, **195**(3), 834–844.
2. H. Nakamura, *Abstract of 62nd Battery Symposium*, 2021.
3. X. Song, J. Pan, L. Xiao, L. Gao and S. Mathur, A hierarchical hybrid design for high performance tin based Li-ion battery anodes, *Nanotechnology*, 2013, **24**(20), 205401.
4. M. Watanabe, *et al.*, Oxidative-Stability Enhancement and Charge Transport Mechanism in Glyme–Lithium Salt Equimolar Complexes, *J. Am. Chem. Soc.*, 2011, **133**, 13121–13129.

5. J. Wu, S. Liu, F. Han, X. Yao and C. Wang, Lithium/Sulfide All-Solid-State Batteries using Sulfide Electrolytes, *Adv. Mater.*, 2021, **33**(6), e2000751.
6. K. Tang, R. J. White, X. Mu, M. Titirici, P. A. Van Aken and J. Maier, Hollow Carbon Nanospheres with a High Rate Capability for Lithium-Based Batteries, *ChemSusChem*, 2012, **5**(2), 400–403.
7. T. Yi, L. Jiang, J. Shu, C. Yue, R. Zhu and H. Qiao, Recent development and application of $Li_4Ti_5O_{12}$ as anode material of lithium ion battery, *J. Phys. Chem. Solids*, 2010, **71**(9), 1236–1242.
8. S. Manabe and R. F. Strickler, Thermal Equilibrium of the Atmosphere with a Convective Adjustment, *J. Atmos. Sci.*, 1964, **21**(4), 364.
9. J. Yan, H. Huang, J. Zhang, Z. Liu and Y. Yang, A study of novel anode material CoS_2 for lithium ion battery, *J. Power Sources*, 2005, **146**(1–2), 264–269.
10. C. Delmas, J. Braconnier, C. Fouassier and P. Hagenmuller, *Solid State Ionics*, 1981, **3–4**, 165.
11. T. Takahashi, *Jp. Pat.*, 2009-135074, 2009.
12. Y. Kuroda, S. Okada, E. Kobayashi and J. Yamaki, *et al.*, *51th Battery Symposium*, *3G12*, 2010.
13. S. Komaba, W. Murata, T. Ishikawa, N. Yabuuchi, T. Ozeki, T. Nakayama, A. Ogata, K. Gotoh and K. Fujiwara, *Adv. Funct. Mater.*, 2011, **21**(20), 3859.
14. S. Cho, D. Y. Kim, J.-I. Lee, J. Kang, H. Lee, G. Kim, D.-H. Seo and S. Park, Highly Reversible Lithium Host Materials for High-Energy-Density Anode-Free Lithium Metal Batteries, *Adv. Funct. Mater.*, 2022, **32**, 2208629.
15. D. A. Stevens and J. R. Dahn, *J. Electrochem. Soc.*, 2000, **147**(4), 1271.
16. A. M. Albatayneh and M. N. Assaf, Comparison of the Overall Energy Efficiency for Internal Combustion Engine Vehicles and Electric Vehicles, *Environ. Clim. Technol.*, 2020, **24**(1), 669–680.
17. P. T. Bhutia, S. Grugeon, A. El Mejdoubi, S. Laruelle and G. Marlair, Safety Aspects of Sodium-Ion Batteries: Prospective Analysis from First Generation Towards More Advanced Systems, *Batteries*, 2024, **10**, 370.
18. H. Nakamura, *Abstract of 65th Battery Symposium*, 2024.